Legitimacy in European Nature Conservation Policy

The International Library of Environmental, Agricultural and Food Ethics

VOLUME 14

LEGITIMACY IN EUROPEAN NATURE CONSERVATION POLICY

Case Studies in Multilevel Governance

Edited by

Jozef Keulartz
Wageningen University,
The Netherlands

and

Gilbert Leistra
Wageningen University,
The Netherlands

 Springer

Editors

Jozef Keulartz
Wageningen University
The Netherlands

Gilbert Leistra
Wageningen University
The Netherlands

Library of Congress Control Number: 2007936991

ISBN: 978-1-4020-6509-5   e-ISBN: 978-1-4020-6510-1

Acknowledgements

This publication was supported by the Netherlands Organisation for Scientific Research (NWO), as part of the programme "Ethics, Research and Governance".

It was also supported by ALTER-net, a 'network of excellence' funded by the EU's 6th Framework Programme.

Contents

Contributors

Vivien Behrens is a junior social scientist at the UFZ-Centre for Environmental Research in Leipzig. She has been working in the European research project FRAP (Framework for biodiversity Reconciliation Action Plans) on biodiversity conversation, analysing conflicts between wildlife management and human resource use. Now, she is also working on social conflicts in the remediation of contaminated sites. E-mail: Viven.behrens@ufz.de

Henk van den Belt is Assistant Professor of Applied Philosophy at Wageningen University & Research Centre. His central interests are science and technology studies (STS), nature policy, bioethics and intellectual property. His thesis *Spirochaetes, Serology and Salvarsan: Ludwig Fleck and the construction of medical knowledge about syphilis* (1997) offers a constructivist analysis of several episodes in the development of biomedical science. He is currently studying the institutionalization of ethics in science policy across Europe and worldwide cultural differences in the socio-ethical evaluation of biotechnology and genomics applications. E-mail: henk.vandenbelt@wur.nl

Susanna Björkell MSc (University of Helsinki) is a researcher at the Centre for administration and organisational, regional and environmental studies, FO-RUM at Helsinki University. She has studied the conflicts associated with the implementation of Natura 2000 in Finland. Local participatory planning of nature conservation areas, new models for nature conservation practices and the use of multiple knowledge sources in legitimising the need for nature conservation are her research interests. E-mail: susanna.bjorkell@helsinki.fi

Dirk Bogaert is a lecturer at the Department of Social Welfare Studies of the Ghent University and at the Arteveldehogeschool Ghent. He is also a senior researcher at the Maritime Institute of the Ghent University. In 2004 he completed his dissertation on nature policy in Flanders with special attention to the institutionalization process, innovation of policy arrangements and processes of multi-level governance. His research projects focus on societal aspects of nature policy and policy processes, the development of indicators, sustainable development education and since 2005 integrated coastal zone management. E-mail: dirk.bogaert@ugent.be

Tadeusz J. Chmielewski is Professor of Landscape Ecology and Nature Conservation in Lublin Agriculture University in Poland. His research interests include: analysis of landscape structure and function (especially peatland and lakeland landscapes); methods and techniques of nature and economy harmonization (in local and regional scale); National & Landscape Parks planning; ecosystem (especially wetland) restoration. His most important publication is two volumes book: *Spatial Planning Harmonizing Nature end Economy*, Lublin 2001. He is designer of the Polesie National Park, the West Polesie Biosphere Reserve, the Roztocze Biosphere Reserve, Ponidzie Region Landscape Parks Cluster and many others protected areas in Poland.

E-mail: tadeusz.chmielewski@ar.lublin.pl, t.chmielewski@pollub.pl

Wim Dubbink is Associate Professor at the Department of Philosophy of Tilburg University. He teaches in the field of business ethics. His field of research is the political theory of the free market. He studied history and philosophy and took a PhD in Public Administration. In the second part of the 1990s he was be employed as Senior Public Official at the Ministry of Agriculture, Nature Conservation and Fisheries (LNV), department Nature Conservation. Alone and together with others he wrote several papers on the Ministry, 'De legitimiteit van het natuurbeleid van LNV.' in *Krisis* (2005).

E-mail: w.dubbink@uvt.nl

Ewald Engelen (University of Amsterdam, Department of Geography, Planning and International Development Studies) was trained as a political philosopher but has since specialized as a financial geographer. In between he has published on corporate governance, democratic theory, the dilemmas of migration, the political economy of immigrant entrepreneurship, the role of pension funds in the financialization of the economy, the deprofessionalization in the public services, and, recently, the local effects of the internationalization of capital. His work has appeared in highly ranking international journals such as Politics & Society, Political Studies, Economy & Society, Environment & Planning A, Journal of Ethnic and Migration Studies, International Migration Review, and Philosophy & Social Criticism.

E-mail: eengelen@fmg.uva.nl

Susana Aguilar Fernández is Senior reader at the Complutense University of Madrid (Spain). 1987 Award of the Spanish Association of Political Science and Sociology to the best minor dissertation. 1994 Letter of commendation of the International Sociological Association at the Second World Contest for Young Sociologists. She has participated in different European Union-financed research projects since 1993. Her most relevant publications in English embrace articles in journals such as The Journal of Public Policy, Environmental Politics, Science and Public Policy, as well as contributions to books published by Routledge, Edward Elgar, Scandinavian University Press, Nomos

Verlagsgesellschaft, Sigma-Berlin, and Kluwer Academic Publishers. She is the author of "El reto del medio ambiente" (The Challenge of the Environment), published in Alianza Editorial, and co-author of "Política Ambiental en España" (Environmental Policy in Spain), published by Tirant lo Blanch.
E-mail: saguilar@cps.ucm.es

Jozef Keulartz is Associate Professor Applied Philosophy at Wageningen University and Research Centre. He has been appointed special chair for Environmental Philosophy at the Faculty of Science, Mathematics and Computing Science of the Radboud University Nijmegen. He has published extensively in different areas of science and technology studies, social and political philosophy, bioethics, environmental ethics and nature policy.

Keulartz is a member of the Scientific Council Large Herbivores (Oostvaardersplassen), set up by the National Forest Service, and of the Scientific Council of The European Centre for Nature Conservation (ECNC). He is a participant in ALTERnet, a 'network of excellence' funded by the EU's 6th Framework Programme. He is also a member of the Netherlands Commission on Genetic Modification (GOGEM).
E-mail: jozef.keulartz@wur.nl

Michiel Korthals is Professor of Applied Philosophy at Wageningen University. He studied Philosophy, Sociology, German and Anthropology at the University of Amsterdam and the Karl Ruprecht Universität in Heidelberg. His academic interests include bioethics and ethical problems concerning food production and environmental issues, deliberative theories, and American Pragmatism. Main publications: *Filosofie en intersubjectiviteit*, Alphen a/d Rijn, 1983; *Duurzaamheid en democratie* (Sustainability and democracy, Boom, 1995); *Philosophy of Development* (Kluwer, 1996 with Wouter van Haaften and Thomas Wren), *Tussen voeding en medicijn* (Between Food and Medicine), Utrecht 2001, *Pragmatist Ethics for a Technological Culture* (with Keulartz et. al.; Kluwer 2002), *Ethics for Life Sciences* (Springer, 2005), *Before Dinner. Philosophy and Ethics of Food* (Springer 2004), and in 2006, *Pépé Grégoire, Een filosofische duiding van zijn beelden/A Philosophical Interpretation of his Sculptures*, Zwolle: Waanders
E-mail: michiel.korthals@wur.nl

Jaroslaw Krogulec is Conservation Director at the Polish partner organization of BirdLife International, the Polish Society for the Protection of Birds. His main task is to secure proper protection of important bird sites in Poland through establishing Special Protection Areas as part of the Natura 2000 network. He graduated from Wroclaw University and obtained his PhD at the Curie-Sklodowska University of Lublin and has since been working there at the Department of Nature Conservation, focusing primarily on the ecology of threatened bird species.
E-mail: Jaroslaw.Krogulec@otop.org.pl

Markus Leibenath is research associate in the department for regional development and landscape ecology at the Leibniz Institute of Ecological and Regional Development (IOER) in Dresden, Germany, as well as lecturer at Dresden

Technical University. His main research interest is in the European dimension of spatial development and in the implementation of biodiversity policies. Recent publications: *Crossing Borders: Natura 2000 in the Light of EU-Enlargement* (with Sandra Rientjes and others, 2005), *Systemic Evaluation of Cross-Border Networks of Actors* (with Robert Knippschild, 2005), and *Greening Europe? Environmental Issues in Spatial Planning Policies and Instruments* (with Karina Pallagst, 2003).
E-mail: M.Leibenath@ioer.de

Gilbert Leistra holds a MSc in forest and nature conservation policy and is currently employed as a PhD-candidate with the Applied Philosophy Group at Wageningen University.

His research focuses on the consequences for the administrative effectiveness, democratic transparency and legitimacy of Dutch nature policy as a result of a double shift in governance: horizontally from the state to the market and to civil society, and vertically from the national level to supra-national and sub-national levels. In his research on the legitimacy of Dutch nature policy he combines his research interests in nature conservation, public policy and political philosophy with his working experience as a policy co-worker at the department for Nature and Landscape of the province Noord-Brabant in the Netherlands.
E-mail: gilbert.Leistra@wur.nl

Pieter Leroy graduated in sociology (Leuven University, 1976) and got a PhD in social and political sciences (Antwerp University, 1983), Belgium. Since 1994 he is Professor of Political Sciences of the Environment at the University of Nijmegen, the Netherlands. He has published on the societal and political aspects of environmental conflicts (nuclear power, the location of industries and infrastructures, the siting of waste facilities etc.), on the strategies of the environmental movement and, mostly, on environmental policies. His recent research focuses on the emergence of new environmental policy arrangements, in the context of more encompassing shifts in governance, and on the analysis and the evaluation of contemporary environmental policies. From 2001 to 2006 he had a part-time professorship at Antwerp University, Belgium, where he co-ordinated a research consortium on environmental policies. He is a member of various advisory boards in Belgium, the Netherlands and France, in particular with regard to the 'state of the environment'-reporting and the role of policy evaluation therein.
E-mail: p.leroy@fm.ru.nl

Elizabeth Oughton is a researcher at the Centre for Rural Economy at Newcastle University. Originally trained as and agricultural and development economist her research has spanned India, Jordan and the countries of central Europe. She now works closer to home on an interdisciplinary project on angling in Yorkshire. Her particular interest lies in the ways in which institutions facilitate and restrict the uses of natural resources and how rural people achieve well-being through their lives and livelihoods. Her publications are in social and feminist economics and in rural sociology.
E-mail: E.A.Oughton@newcastle.ac.uk

Florence Pinton is associate Professor (Maître de conférence) in Sociology at the University of Nanterre (Paris 10) and she has been working at the Institute of Recherche for the development (IRD, Orléans) between 2003 and 2007 where she is member of the "Environment policies team". Her research fields include territorial, social and political changes induced by new generation of nature policies which aspired to sustainable development. She works in France and Brazil. During 6 years, she was the scientific coordinator of a research program on the implementation of Natura 2000 in France supported by the Ministry of Environment and the french Institute of Biodiversity (IFB). She has edited a book include the analysis of the directive Habitats in a contexte of decentralization in decisions: " *La construction du réseau Natura 2000 en France: une politique publique à l'épreuve des scènes locales*" (Documentation française).
E-mail: florence.pinton@orleans.ird.fr

Felix Rauschmayer is senior research fellow at the Helmholtz Centre for Environmental Research UFZ. He works and publishes in decision analysis, participation, environmental governance and biodiversity conservation. He is author of *Entscheidungsverfahren im Naturschutz: Die Multikriterienanalyse* (1999) [Decision procedures in nature conservation: the multi-criteria analysis]. Recently, he was guest-editor for Land Use Policy for a special issue on resolving environmental conflicts. Currently, he is co-editing *Human Wildlife Conflicts in Europe—Fisheries and Fish-eating Vertebrates as a Model Case*.
E-mail: felix.rauschmayer@ufz.de

Jac A.A. Swart is Assistant Professor in the Science & Society Group (SSG) of the Faculty of Mathematics & Natural Sciences, University of Groningen, the Netherlands. Trained as a biologist, he later moved into the humanities. His academic interests include the societal, ethical and dynamic aspects of life sciences. He has contributed to journals in the fields of natural sciences and philosophy on pharmacodynamics, conservation ecology, knowledge systems, societal and ethical aspects of nature conservation and sustainability, and animal ethics. His current research topics include food genomics and biotechnology in Third World countries.
E-mail: j.a.a.swart@rug.nl

Jan van Tatenhove is Associate Professor at the Environmental Policy Group at the Department of Social Sciences at Wageningen University & Research Centre. His research focuses on (1) processes of political modernisation and the innovation of policy arrangements in several policy domains, such as environmental policy, nature conservation and water management (2) the institutionalization of new political spaces, power and interactive policymaking, and (3) processes of multilevel governance in EU-policy making and the interplay of formal and informal practices.
E-mail: jan.vantatenhove@wur.nl

Jane Wheelock is Professor of Socio-economics at Newcastle University, where she has worked since 1992. She has been researching household responses to

economic change since 1985. She has specific interests in the impact of economic insecurity and environmental change on household provisioning, with a recent focus on rural economic development implications. Most recently, she has been developing an interdisciplinary approach to social reproduction and household well being. Key publications include: *Insecure Times: Living with Insecurity in Modern Society*, co-edited with John Vail and Michael Hill (Routledge, 1999) and *Work and Idleness: the Political Economy of Full Employment*, co-edited with John Vail (Kluwer, 1998).
E-mail: Jane.Wheelock@ncl.ac.uk

Henny van der Windt is Assistant Professor in the Science & Society Group (SSG) of the Faculty of Mathematics & Natural Sciences, University of Groningen, the Netherlands. He studied biology and gained his PhD in natural sciences. His fields of interest include the dynamics of science, science policy, social movements and environmental history. His main publications concern the development of ecology, nature conservation, ecological standard setting and visions on nature. His current research topics include food genomics, science communication and bionanotechnology.
E-mail: h.j.van.der.windt@rug.nl

Rüdiger K.W. Wurzel is Reader and Jean Monnet Chair in European Union Studies at the University of Hull (United Kingdom) where he is Director of the Centre of European Union Studies (CEUS). He has published widely on European Union and member state environmental policies and other public policy issues. Rüdiger Wurzel has carried out a large number of research grant funded projects and is a member of the editorial board of the Journal of European Public Policy. Recent publications include Environmental Policy-Making in Britain, Germany and the European Union. The Europeanisation of Air and Water Pollution Control, Manchester University Press (2006), The EU Presidency: 'Honest broker' or driving seat? An Anglo-German Comparison in the Environmental Field, Anglo-German Foundation (2004) and 'New' Instruments of Environmental Governance? National Experiences and Prospects, Frank Cass which is jointed edited with A. Jordan and A. Zito (2003).
E-mail: R.K.Wurzel@hull.ac.uk

Erik van Zadelhoff currently works for IUCN, the World Conservation Union in Brussels and Amsterdam. He works on the conservation and sustainable use of global ecosystems in relation to European activities. Until recently he was head of the Department for Nature and Environment of the National Reference Center for Agriculture, Nature and Food Quality of the Ministry of LNV. Erik was secretary of the working group that developed the first Nature Policy Plan of LNV between 1987 and 1990. Next to numerous publications on ecological topics, he published several articles on the development of the nature policy in the Netherlands.
E-mail: f.vanzadelhoff@chello.nl

Marielle van der Zouwen is assistant professor at the Forest and Nature Conservation Policy Group, Department of Environmental Sciences at Wageningen University, the Netherlands. She recently completed her dissertation, entitled 'Nature policy between trends and traditions – dynamics in nature policy arrangement in the Yorkshire Dales, Doñana and the Veluwe' (2006) in the field of political sciences of the environment. Her research projects focus on change and stabilisation in (multi-level) governance practices in the EU and on the relation between traditional and new policy arrangements in these practices, specifically in the field of environmental, nature, forest and rural area policy.
E-mail: marielle.vanderzouwen@wur.nl

Hub Zwart is Full Professor of Philosophy, Faculty of Science, Radboud University Nijmegen, the Netherlands; Department of Philosophy & Science Studies (Chair); Centre for Society & Genomics (Director); Institute for Science, Innovation & Society (Director). He studied philosophy and psychology at the Radboud University Nijmegen, worked as a research associate at the *Institute for Bioethics* in Maastricht, defended his thesis in 1993, was subsequently appointed as research director of the *Center for Ethics* at the Radboud University Nijmegen, is now full professor of philosophy at the *Faculty of Science*. His main field of interest is philosophy and ethics of the life sciences, notably genomics, with a special interest in science and imagination (comparative epistemology). He has published on the epistemology and ethics of animal research, on the past, present and future of scientific publishing and on environmental policy and the philosophy of landscape. He was European Lead of the International Student Exchange Program *Coastal Inquiries*.
E-mail: h.zwart@science.ru.nl

Part I
Introduction

Chapter 1
European Nature Conservation Policy Making

From Substantive to Procedural Sources of Legitimacy

Ewald Engelen, Jozef Keulartz and Gilbert Leistra

1.1 Introduction

Europe is not only rich in history and culture but also in nature. Its natural diversity, often intensely shaped, moulded and influenced by human action, has enormous ecological, aesthetical and economic value. Due to the continuing deterioration of natural habitats and the threats posed to certain species, preservation, protection and improvement of the quality of the environment, including the conservation of natural habitats and of wild fauna and flora, have increasingly become objectives of environmental policy-making at the level of the European Union (EU). Emblematic of this newly felt responsibility is the fairly recent initiative 'Natura 2000', which aims at creating a coherent network to protect the most valuable nature areas in the EU. The creation of this network is, without doubt, one of the most ambitious supra-national initiatives for nature conservation worldwide and forms the cornerstone of the EU nature conservation policies.

The idea of Natura 2000 was first broached in the Council Directive 92/43/EEC of 21 May 1992 on the conservation of natural habitats and of wild fauna and flora. This Habitat Directive stated that in order to ensure the restoration or maintenance of natural habitats and species that are of interest to the Community, it was necessary to designate Special Areas of Conservation (SACs) to create a coherent European ecological network. Apart from that, it was decided that the Natura 2000 network would also incorporate Special Protection Areas (SPAs), whose designation would be up to the member states in accordance with Council Directive 79/409/EEC, better known as the Birds Directive.

The Birds Directive, the precursor of the Natura 2000 initiative, was the first piece of EU legislation in the field of nature conservation. It reflected international indignation concerning the killing of migratory birds, especially in the Mediterranean area, which took such extreme forms that the survival of these birds was

Ewald Engelen
Department of Geography, Planning and International Development Studies,
University of Amsterdam
E-mail: eengelen@fmg.uva.nl

endangered. Research had demonstrated that more than half of the bird species in Europe were in decline. Although the Directive was an uncomfortable compromise between the ecological value of protection and the economic value of exploitation, it was still perceived as a great step forward (Osieck 1998); not only was it the first action of the EU in this field, but it also forced all member states to commit themselves to protect more than 200 endangered bird species as well as the areas that were crucial to their survival.

The Natura 2000 network and the directives of which it consists must be seen in the broader context of the international commitments that the EU and its member states have taken up. The most important of these are the 1992 Convention on Biodiversity (CBD), the 1971 Ramsar Convention on Wetlands and the 1979 pan-European Convention on the Conservation of European Wildlife and Natural Habitats (Bern Convention). The 1992 Convention on Biodiversity (CBD), signed by 150 government leaders at the 1992 Rio Earth Summit and dedicated to promoting sustainable development, resulted in the EU's commitment to stop the decline of biodiversity within its territory by 2010. The Convention on Wetlands, signed in Ramsar, Iran, in 1971, is an intergovernmental treaty, which provides the framework for national action plans as well as international cooperation, aimed at the conservation and sustainable use of wetlands and their resources. The Bern Convention launched the idea of a pan-European network of protected habitats, known as the Emerald Network. The Bern Convention and Natura 2000 have similar objectives, since both are international legal instruments aimed at the conservation of wild flora, fauna and natural habitats. Their main difference concerns the territory to which they apply. The Emerald Network envisages a network of Areas of Special Conservation Interest (ASCIs) in Europe and parts of Africa, while Natura 2000 focuses only on the EU member states.

Nature 2000, which encompasses both the Habitats and Birds Directives, is the most important legal framework in the field of nature conservation in Europe. These directives are legally binding texts that require the individual member states to translate them into national legislation. As with other EU directives, they give the member states a significant amount of leeway as to the exact rules to be adopted. If the member state fails to pass the required national legislation, or if the national legislation does not adequately comply with the requirements of the directive, the European Commission can initiate legal action through the European Court of Justice (ECJ).

Although significant successes have been achieved, reflecting the largely uncontested nature of the overall goals of Natura 2000, we can, nevertheless, observe increasing implementation problems in the different EU member states. The difficulties and delays that have beset implementation in the 'old' 15 EU member states (i.e. states granted EU membership before May 2004) have resulted in several member states – including Denmark, Finland, France, Germany, Greece, Ireland, and the Netherlands – being taken before the ECJ by the European Commission for failure to submit lists of designated sites in accordance with the Habitats Directive (Paavola 2004).

There are several reasons why the implementation throughout the EU causes increasing problems. The purely scientific legitimation underlying Natura 2000 has proven to be both a strength and a weakness. To its benefit, Natura 2000 can claim strong support from ecological experts in other supranational and national conservation organizations. The World Wildlife Fund (WWF), for example, 'strongly support[s]' the establishment of Natura 2000, as it is perceived to be a 'real attempt to conserve Europe's biodiversity [...]. The sites will not just be a collection of national or regional parks designated for a variety of reasons' (De Pous and Beckmann 2005: 3). However, the scientific legitimation of Natura 2000 is also a source of weakness because the scientific ecological concepts underlying the Natura 2000 network seem to invite a technocratic, top-down mode of policy-making that is increasingly being rejected by local constituencies as too insensitive to local interests, too paternalistic for modern tastes and too elitist for modern democracies. Low political priority and a tight implementation schedule have, in many cases, made the implementation process even more problematic, forcing governments to speed up the process and limit the extent to which local interests can collectively decide on the implementation process (Paavola 2004). Lack of communication, information and recognition has further added to the national and local resistance to the implementation of Natura 2000. In France, for example, the lack of communication outraged local stakeholders, who questioned both the science-based site designation as well as the quality of the information on which the designations were based (Alphandéry and Fortier 2001). Of a different nature are the challenges faced by the new member states. Many of the Natura 2000 sites proposed in the accession countries have been under threat, not only because EU nature conservation policies are poorly integrated with other sectoral policy-making trajectories, but also because the available resources and regulatory capacities are limited (De Pous and Beckmann 2005).

In other words, the EU's evident legislative success in the field of nature conservation policies upstream has increasingly been frustrated by highly politicized decision-making downstream, which has resulted in delays, deadlocks, and even out-and-out resistance. In response, member states have increasingly resorted to more participatory forms of decision-making. Public consultation, interactive governance, and deliberative democracy are some of the key models that are currently being used in many member states to overcome opposition, enhance the legitimacy of policy outcomes, increase the quality of decision-making and facilitate implementation.

It is this state of affairs that is the main focus of this volume. We have invited a number of experts on nature conservation policies from different EU member states – old as well as new members – to reflect upon changes over time in the way in which 'legitimacy' is being 'produced' within the country that is their main object of study. The expectation underlying this volume is that, in most member states, we should observe a shift from substantive sources of legitimacy to more procedural ones, implying a shift from vertical policy-making by experts to a more horizontal mode of policy-making through democratic deliberation. The causes behind this shift are easy to identify. First, there is the increasing politicization of nature conservation policies as soon as they hit the ground. What is easily perceived as a mere technical issue upstream appears to have strong distributive consequences downstream. Since most EU directives in the field of nature policy-making have reached the downstream phase of the policy trajectory, we should expect increasing politicization and, hence, an urgently felt need on the part of policy-makers for new modes of legitimacy 'production'. The other cause behind the shift from substantive to procedural sources of legitimacy has to do with the increasing delegitimation of scientific expertise, as a result of both the post-modern relativization of scientific truth claims and the increasing politicization of scientific prescriptions.

Before giving an overview of the content of this volume, we will first present the framework from which we have derived these expectations as well as the research questions that we aim to answer in this volume.

1.2 Theoretical Framework

1.2.1 Shifts in Governance

Keying the word 'governance' into Google generates well over 200 million hits, most of them from the 1990s and 2000s. This highlights both the relative novelty as well as the increasing popularity of the concept. While no comprehensive history of the concept has been written yet, it is obvious that its rapid dissemination within the social sciences is related to real world changes that can be traced to the momentous macro-economic shifts of the 1970s. The neo-liberal project of market-making that was (enthusiastically or reluctantly) pursued by a growing number of national political and economic élites as an answer to the predicaments of economic stagnation, growing unemployment and rising inflation, raised the spectre of the end of the nation-state as we know it (Strange 1996; Ohmae 1995). While the expectation of a stateless world proved to be too pessimistic or optimistic (depending on the perspective), what has come about, especially in the European region, is a regulatory order in which the state and national democratic institutions have come to play other roles from those they used to play. States now have to share their regulatory powers with a growing number of multinational and supranational organizations and institutes, while being forced by growing societal differentiation and fragmentation to pursue policies of federalization, devolution and subsidiarity that have enhanced the importance of local polities and local agents. At the same time, due to the market-making and market-extending effects of neo-liberal globalization, a growing number of regulatory and decision-making tasks that used to be the prerogatives of the Westphalian state have either intentionally or unintentionally been taken over by private agents such as firms, labour unions, employers organizations and other non-governmental agencies.

Since the social sciences are by nature backward looking, there was a certain time lag before these real world developments actually started to change the conceptual map of disciplines like political economy, sociology, public administration, international relations and political science. Since the 'shifts' described above touch upon almost every dimension of social life – ranging from procreation to migration – almost every discipline has been confronted with the need for conceptual retooling. Cherished concepts like power, labour, community, society and economy had to be cleansed from their methodological nationalist overtones (Smith 1983; Glick Schiller and Wimmer 2003). The same was true for the concept of 'government' and *mutatis mutandis* the concept of 'state'. Since the conceptual history of 'governance' is actually a multitude of 'local' histories, it is hard to capture in one general tale (see also Van Kersbergen and van Waarden 2001). Nevertheless, three observations can be made that are at the core of the conceptual renewals that were forced upon the social sciences by the developments noted above.

The first is the *demystification of human rationality*. Functionalism, in each and every shade, presumes an orderly world that is knowable and changeable by rational agents using scientific insights that are neutral and objective. As such, societal changes are supposed to culminate in the one best way of doing things that would deliver optimal outcomes. If the crises of the 1970s demonstrated anything, it was that human knowledge had proven to be inherently fallible, limited and ideologically blinkered. This 'discovery' not only marked the end of modernization theory, which had proclaimed the ultimate triumph of liberal democracies and capitalist economies (Kerr 1960; Parsons 1966), but it also initiated a growing reflexivity among scientists as well as policy-makers concerning the nature of the knowledge on which they based their interventions. For limits of rationality implied limits of changeability, as was demonstrated by a growing number of large scale planning disasters (Scott 1998), and hence forced policy makers to look for other instruments to solve societal problems than the top–down approaches of yesteryear that presupposed infallible knowledge. Here, the post-modern critique of empiricist and positivist conceptualizations of scientific knowledge joined hands with the new left wing critique of undemocratic and paternalist modes of top–down planning of the 1960s and 1970s and the older conservative critique of rationalism per se, which resulted in an increased 'politicization of scientific expertise' as an answer to the 'scientification of politics' preceding it (Gouldner 1970; Rueschemeyer and Skocpol 1996; Oakeshott 1962).

Second, the discovery that states are limited in what they can do and that they are highly dependent on other agents for agenda setting, legitimate decision-making and effective implementation led social scientists in the 1980s to 'discover' that states are not godlike 'unmoved movers' but that they act in a 'shared public space' (Crouch and Streeck 1993). Around that time, it also became increasingly accepted that the dispersal of sovereignty this discovery implied was not something new, but had, in fact, always characterized each and every state to some degree, raising pressing questions about continuity and change. Nevertheless, a growing conceptual need to decouple state and government arose since non-state agents increasingly took over tasks and responsibilities that could easily be called 'government' as the concept is understood in the traditional sense. In other words, state and government were no longer identical, raising the question of how to refer to that which non-state agents were doing? And what did the actual distribution of governing tasks imply for the dominant conceptualization of the nation-state as being 'sovereign' within its own territorial jurisdiction?

The concept of 'governance' dates back to the late fourteenth century; from a period that is of well before the rise of the nation state (OED 1989). It survived that rise as a concept denoting forms of regulation that closely resembled state regulation – in form as well as in substance – but which were undertaken by agents that did not possess the defining properties of the nation-state, i.e. a monopoly on the use of violence, minting money and raising taxes (Weber 1972: 29). Examples of such nongovernmental modes of governance were colonial governance or the governance of guild-like institutes like universities and churches. After a long period of near total neglect, the concept was 'rediscovered' by political scientists and students of public administration in the early 1980s and was rapidly taken up to fill this conceptual gap. Here, the older literature on corporatism as well as the more prospective subdiscipline of public administration proved instrumental (Schmitter 1974; Streeck and Schmitter 1985; Rhodes 1997; Pierre and Guy Peters 2000; Pierre 2000; Kjær 2004).

Finally, building upon both the work of German scholars investigating the functionality and dysfunctionality of federalism and the work of international relations-scholars investigating the problems of delegation in the case of supranational bargaining and negotiation, called 'two level games' (Scharpf 1988; Benz et al. 1992; Putnam 1988), an increasing number of students of international law, international relations and international political economy, as well as political science more broadly, started to couch the phenomenon of European integration in terms of multilevel governance (Benz 1998; 2000; Benz and Eberlein 2001; Scharpf 1999; 2003; Hooghe and Marks 2001; 2004).

In a recent review paper, the American political scientists Gary Marks and Liesbeth Hooghe have identified two 'contrasting visions of multi-level governance' in the EU and beyond (Marks and Hooghe 2004). The first type of governance system retains the territorial basis of the Westphalian state as well as its general-purpose orientation, but distributes jurisdiction, responsibility and power over a limited, well-defined number of subnational and supranational territorial levels, i.e. the local, the national and the supranational or regional level. In one normative vision of the EU's future, European citizens will become subjects of three, well-defined territorial polities simultaneously. The other type of system is less easy to qualify because it dispenses with the transparency of territorial organization. Instead, in this vision, multi-level governance systems are being conceptualized as functionally based, overlapping and temporary. In this explicitly post-nationalist perspective, problem solving is the dominant orientation, while 'exit' is its dominant virtue. Both models have different fields of application. While the first type, in one form or other, predominates in the seemingly endless discussions on how to characterize the EU, scholars more interested in other fields of supranational decision-making tend to lean towards the second model. Despite these differences, though, both models raise the problem of legitimacy, since the predominant mode of legitimacy production through territorially based institutions of representative democracy are clearly insufficient, albeit in different degrees, for legitimacy production within the two types of governance regimes.

In summary, both the 'discovery' of the limits of rationality, the replacement of government by 'governance' and the multiplication of levels and places of policy-making point to the empirical and theoretical need to find new modes of legitimacy production (see Scharpf 2004). The state is no longer omnipotent and omniscient, has lost its monopoly on policy making to other public and private agents and has increasingly become embedded in a multi-level governance structure that combines territorial and functional modes of organization in a complex and incoherent way. This state of affairs increasingly cries out for new modes of democratic legitimacy. However, what is legitimacy and why is it needed? The next subsection addresses these questions

1.2.2 The Problem of Legitimacy

Generally speaking, legitimacy refers to the question of why the outcomes of binding collective decision-making ought to be accepted by those whose interests are being harmed by the decision in question (Scharpf 1970; 1999; 2004; Weber 1972: 16–17, Ch. 9). While seemingly an intrinsic element of all collective decision-making, the issue of legitimacy, in fact, arises only under very specific conditions, which we will call the *circumstances of legitimacy*, analogous to David Hume's famous analysis of the conditions under which issues of justice arise.

The first one has to do with the presence of a political system in which individual interests are seen as legitimate concerns that cannot be harmed without valid reasons. In political systems in which individual interests do not serve as self-standing concerns that might clash with collective imperatives, legitimacy is not an issue. In fact, the hallmark of totalitarian regimes is precisely the absence of any distinction between collective and individual interests. In such a case, consent is either based on coercion (as in national-socialist Germany), on tradition (as in absolutist monarchies), on religion (as in theocracies such as present-day Iran) or on scientific expertise (as in the former Soviet Union). Despite the fact that the subordination of individual interests to collective ones is not an exclusively pre-modern phenomenon, as is demonstrated by the examples of Germany and the Soviet Union, we call this the *circumstance of modernity*.

Second, for the issue of legitimacy to arise, individuals have to be *interdependent*. In a world of perfectly autonomous individuals, there is simply no need to make binding collective decisions, for there is no higher order commonality (the public) whose interests could collide with those of the constituting individuals. It is easy to recognize standard social contract-theory in this line of argumentation. The more the 'state of nature' is conceptualized as a zero-sum game – strongly in the case of Hobbes, less in the case of Locke, and not at all in the case of Rousseau (Hampsher-Monk 1992) – the stronger are the legitimacy requirements for the ensuing social contract. Basically, this is the reverse of the first circumstance, namely the recognition of individual interests over and above those of the collective as legitimate concerns in collective decision-making. Here, at the opposite side of the continuum, the issue is whether there is actually a collective concern over and above individual interests, which requires collective decision-making. We call this the *circumstance of interdependence*.

The third circumstance has to do with the nature of the issue that is at stake. As the German sociologist Fritz Scharpf has argued, the legitimacy requirement of different types of collective decisions is strongly related to their contentiousness (Scharpf 2004). Least contestable, according to Scharpf, are so-called 'pure coordination' games. Since all participants have an equal interest in solving coordination problems and are indifferent to the actual solution, the decision does not transgress individual interests and is hence not in need of legitimacy. Much stronger are legitimacy requirements in the case of so-called 'dilemma' games, like the famous prisoners' dilemma, in which the binding force of collective courses of action is severely weakened by the availability of opportunistic choice options. While the collective benefit is maximized if all agents pursue the same course of action, individual benefits are highest if the individual agent reneges on his promises. In order to minimize opportunistic behaviour, a credible 'shadow of hierarchy' is required (Scharpf 1997), which is itself in need of legitimation. The strongest requirements pertain to so-called 'zero-sum' games. Each and every decision in such a context implies a redistribution of resources from some agents to others. While in the other two games it can be argued that the decision ultimately serves the enlightened self-interest conditions, even of the losers, no such strategy is available in the case of zero-sum games.

The fourth circumstance refers to the *source of legitimacy*. To make binding collective decisions palatable to individuals whose interests are being harmed, decision makers can choose either *substantive* or *procedural sources of legitimacy*. In democratic political systems, which are based on modern conceptions of individual autonomy and equal moral worth, consent cannot be based on coercion nor can it be derived from tradition or religion. Given the 'condition of pluralism' (Rawls 1995), there is no longer a shared conception of the good life that could serve as a substantial base for legitimate collective decision-making. As a result, legitimacy increasingly has to be produced procedurally. Democracy is precisely that: a procedural solution to the inability of finding broadly acceptable sources of substantial legitimacy, which is not to say that it is only that. Obviously, a similar line of reasoning can be made with regard to scientific expertise as a substantial source of legitimacy. The more there is controversy over the truth-value of the scientific theories underlying public interventions, the more there is reason to 'politicize' the decision in question and to resort to procedural means of legitimacy production. In this regard, there is a strong link between the recognition of the limits of rationality that was seen above as being one of the main drivers behind the three 'shifts in governance' and the increasing saliency of procedural means of legitimacy production.

Since *substantive* sources of legitimacy – religion, charisma, tradition, scientific expertise (Weber 1972: Ch. 9.; Friedrich 1972) – have increasingly lost their legitimating force, *procedural* modes of legitimacy production have willy-nilly gained in importance. Procedural legitimacy can be produced in one of three ways: (i) by letting agents decide for themselves, (ii) by ensuring that collective decisions serve individual interest, or (iii) by guaranteeing a high level of fairness in the decision procedure itself. The first type of procedural legitimacy is known as *input legitimacy*, the second as *output legitimacy* and the third as *throughput legitimacy*. What is determining the degree of input legitimacy is the extent to which those that are subject to a collective decision can (co)determine the agenda. While strongly rooted in the tradition of direct or participatory democracy (see Scharpf 1971; 1999), input legitimacy in current debates is generally operationalised as fair, inclusive and encompassing agenda-setting procedures. In the case of output legitimacy, the emphasis is instead on the extent to which the outcomes of the decision-making process serve the 'objective' needs and interests of the subjects in the decision in question. Whether the subjects had any say over the input-side of the policy cycle is less important. This type of legitimacy is part and parcel of the theoretical tradition of 'democratic minimalism', which stresses the dangers of majority rule and emphasizes the need for safety mechanisms such as the 'trias politicas', federalism and two-chamber decision-making (see Barber 1984). According to Scharpf, the legitimacy of the EU is largely of this type (1999). Throughput criteria, finally, have mostly been formulated in the body of literature that has become known as 'deliberative democracy' theory, a mostly normative approach to issues of legitimacy in contemporary democracies, which emphasizes the importance of the design of the actual decision-making procedure – ensuring fair and inclusive fora in which each participant has equal standing and equal speaking time – for the transformation of individual *interests* into collective *reasons* (Elster 1998; Bohman 1998).

1.2.3 Research Questions

The research questions that we try to answer in this volume are descriptive, explanatory as well as prescriptive. First of all, we are interested in the question whether the Europeanization of national nature conservation policy-making has actually resulted in the multi-agents and multi-level policy setting of the sort identified by Hooghe and Marks. In particular, we want to know whether it is a setting largely resembling the fairly transparent and static territorial model of the nation-state or whether it looks more like what Philippe Schmitter calls a 'consortio' or a 'condominio', i.e. a 'postsovereign, pretero-national, multilayered, overlapping, indefinite arrangement of authority' (Schmitter 2000: 17).

Second, we want to find out what consequences the coming about of such a complex regulatory framework has had on traditional modes of legitimacy production. Can we indeed perceive a shift from substantive to procedural sources and within the latter from output legitimacy to input and throughput legitimacy? Is this a development that can be perceived throughout the EU or only within the 'old' member states? If the latter, why? And what does this tell us about the conditions under which legitimacy issues arise? Can the game-theory explanations offered above actually account for the empirical differences in the urgency of legitimacy issues across the European region?

Finally, we focus on alternative institutional arrangements for the traditional modes of democratic legitimacy production. In a context of widely distributed regulatory prerogatives and democratic responsibilities, which, moreover, appear to have become increasingly mobile rather than fixed and static, the models of democratic legitimacy of yesteryear have become increasingly obsolete. Instead, we need to look for alternative arrangements that fit this new reality while providing at least a modicum of democratic accountability, without standing in the way of effective decision-making. While a prospective, normative exercise that clearly fits the discipline of political philosophy, we contend that praxis is more intelligent than theory with regard to institutional design. Hence, we aim to derive from the case studies presented in this volume some practical suggestions on how democratic legitimacy could be produced in the new regulatory reality this volume depicts.

1.3 Outline of the Book

In this volume, nine case studies are presented: two case studies on protected species (geese in the Netherlands, and the great cormorant in Denmark, France and Italy), four case studies that zoom in on specific protected areas (in Spain, Finland, Poland and the UK), and three case studies with a focus on the implementation of Natura 2000 at the country level (Belgium, France and Germany). These case studies are followed by extensive comments. The volume concludes with a chapter that situates these contributions within the wider EU environmental policy and political context and that reflects on the main themes of the book, which include legitimacy, policy (or state) failure, trust, scientific versus local knowledge and bargaining versus deliberation.

1.3.1 Protected Species

Gilbert Leistra, Jozef Keulartz and Ewald Engelen open this part with a study on wintering geese in the Netherlands. Scientific and ecological expertise has been the starting point of Dutch nature conservation policy. These policies often conflicted with the interests of local stakeholders and generated much resistance. In response, governmental authorities gradually abandoned their centralist, top-down approach and increasingly switched to methods of participatory and interactive policy-making. Substantive sources of legitimacy, in particular scientific expertise, have been replaced by more procedural forms of legitimacy production through public participation processes. This implies that legitimacy can no longer be assumed at the output-side of the decision-making cycle only, but requires production throughout the entire policy process.

The case study of wintering geese in the Netherlands provides an excellent opportunity to gain insight into the problems and possibilities of legitimacy production in what has clearly become a highly controversial arena of policy making and policy implementation. This case study illustrates the shift from substantive to procedural legitimacy production and highlights the dilemmas that can be encountered when making trade-offs between the legitimacy requirements in the input, throughput, and output phases of the policy process.

Wim Dubbink questions most of the empirical claims that are supposed to support the story of the wintering geese. Notwithstanding its top-down character, Dutch nature policy has enjoyed a considerable degree of legitimacy. The Nature Policy Plan from 1990 met with great public support and was backed up by all political denominations. Dutch nature policy was also rather successful. The process of implementation of the Nature Policy Plan is still on schedule, and this has been achieved without the use of judicial power. Dubbink also questions the supposed transformation of Dutch nature policy from top–down to bottom–up planning. Whereas the policy formation process can rightly be called a clear example of top-down planning, the implementation process cannot be characterised in this way due to the strong corporatist tradition in The Netherlands.

Dubbink also questions the moral and political claims underpinning the chapter on wintering geese. Above all, he criticizes the alleged identification of legitimacy with the acceptability of a policy to those directly affected by that policy. Not unlike Henk van den Belt in this volume (chapter 17), Dubbink alerts us to the danger that too much stress on interactive policy will make politicians and policy-makers hostage to local interests at the expense of broader interests, in particular, the interests of future generations in a healthy and beautiful natural environment (see also Fernandez, chapter 6 in this volume).

Felix Rauschmayer and Vivien Behrens discuss the case of the great cormorant in the EU. Through persecution and the destruction of habitats, the numbers of cormorants fell to a very low level all over Europe in the 1960s and the species was threatened with extinction. In 1979, the EC Birds Directive protecting wild birds was enacted. This protection status as well as the protection of breeding sites and the available food supply have rapidly led to increased cormorant numbers since the 1980s. This conservation success posed a threat to certain fish species and led to conflicts with fishers and anglers organisations. As a result of this escalating conflict between species protection and fisheries, Rauschmayer and Behrens observe a shift from species preservation to species management. The purpose of species management is to maintain a minimum viable population and at the same time to safeguard conflicting economic and cultural interests.

According to Rauschmayer and Behrens, the shift from species preservation to species management poses new challenges to the question of legitimacy, especially in a multi-level governance setting. They examine the changes in the production of legitimacy with respect to the criteria of legal compatibility, accountability, interest representation and transparency.

In his comment on the chapter on the great cormorant, Hub Zwart raises the issue of perceptions or visions of nature, an issue that will also be addressed by Jacques Swart and Michiel Korthals in this volume (chapters 9 and 19). According to Zwart, Rauschmayer and Behrens' story of the great cormorant attests to the emergence of an influential new view on nature, the 'governance' view. Whereas the 'rationalist' view perceives nature as the domain of the non-human, the non-moral and the non-rational, and the 'romantic' view envisions nature as a realm of symbiosis, equilibrium and harmony, the governance view sees nature as an arena where interests and claims of various stakeholders, i.e. species, including our own, may conflict. To manage inter-species tensions and conflicts, we cannot simply impose our rational laws, nor can we rely on nature's own resources for self-governance, but we must enter into a complicated process of negotiation and deliberation and of monitoring and evaluating, in which the claims and interests of all species involved must be taken into account. Zwart gives a brief sketch of the way the governance view on nature frames the new mode of knowledge and legitimacy production.

1.3.2 Protected Areas

This part opens with a chapter by Susana Aguilar Fernandez on the establishment of nature conservancy policy in Spain, especially in Doñana in southern Spain (Andalusia), a wetland of international importance. The first initiatives to protect Doñana in the late 1960s were clearly top-down. Under the undemocratic Franco regime there was a lack of input legitimacy – Aguilar Fernandez characterizes the nature conservation policy in Doñana as 'enlightened despotism' and 'ecological dictatorship'. This changed during the democratic transition in 1977 when a decentralization process set in, adding a new level of decision-making. Yet another level was added in 1986 with Spain's accession to the EU.

Aguilar Fernandez notes that decentralization went hand in hand with the hollowing out of the state's role in nature conservation policy and with serious coordination problems between national and regional authorities. At the same time the emphasis shifted to input and throughput legitimacy. This shift took place at the expense of output legitimacy. In an atmosphere of 'building fever', pro-development interests that formed a powerful alliance with municipal authorities, farmers and cattle-breeders, were able to severely hamper the effectiveness of nature conservation policy. The main lesson learned from this experience, Fernandez maintains, is that broad social participation (input legitimacy) and even impeccable environmental governance (throughput legitimacy) do not necessarily produce good policies (output legitimacy). On the other hand, under today's democratic conditions something like an 'enlightened despotism' cannot in fairness be advocated.

According to Jan van Tatenhove, Aguilar Fernandez does not provide a solution to the policy paradox that she frames as the tension between output legitimacy, on the one hand, and input and throughput legitimacy, on the other. Van Tatenhove challenges Fernandez's assumption that input and throughput legitimacy are not compatible with output legitimacy in realizing sustainable nature policy. For a better understanding of the relation between the various kinds of legitimacy in the multi-level context of European governance, we should analyse in detail the specific 'policy arrangements' that have been developed in a certain area. These arrangements include discourses, actors and coalitions, power and resources, and rules of the game (see Van der Zouwen, chapter 13 in this volume). Also needed are detailed analyses of the dynamic interplay of formal and informal practices at the different stages of the decision-making process in the EU.

Susanna Björkell draws on the case of Bergö-Malax outer archipelago to illustrate the search for new participatory models in nature resource management in Finland after Natura 2000. During the implementation of Natura 2000 in Finland, trust in the environmental administration among landowners and local actors was severely undermined due to lack of information and accountability. Today, the need for re-establishing trust has been recognized and deliberation and local participation are seen as crucial to successful nature conservation planning.

Björkell uses theories of co-management and common-pool resource management to investigate the factors that determine success or failure of nature conservation policies, such as taking into account existing community structures, and enabling local access to and influence on the different natural resources management rules.

In his comments on Björkell, Jacques Swart draws attention to the important role that visions of nature play in legitimizing nature policy. Like Hub Zwart and Michiel Korthals in this volume, Swart stresses the need to analyse nature management conflicts in terms of the different and often diverging perceptions or visions of nature that the various stakeholders have. He identifies three visions of nature, the functional vision, the arcadian vision and the wilderness vision, each of which can be characterized with respect to their underlying ecological, ethical and aesthetic outlook.

To understand the emergence, magnitude and course of nature management conflicts, we should inquire into the relationship between these multiple visions of nature, on the one hand, and the multiple spatial and temporal scales that are relevant in nature policy, on the other. Swart offers a detailed discussion of the kinds of conflicts that may arise in the case of discordance between these scales, the concomitant knowledge traditions and levels of decision-making.

Tadeusz J. Chmielewski and Jaroslaw Krogulec discuss the case of the UNESCO West Polesie Biosphere Reserve as an example of a successful bottom–up nature conservation policy in Poland. As a new EU member state, Poland had to designate Special Protection Areas (SPA's) under the Birds Directive and to present a complete national list of proposed Sites of Community Importance (pSCI's) under the Habitats Directive by 1 May 2004 – the day that Poland and nine other countries joined the EU. The designation process started in 2000. The preliminary list of Natura 2000 areas covered about 20 per cent of the country's area. However, due to severe criticisms of the designation process by local authorities, foresters and water management institutions, the final national list, prepared for submission to the European Commission, was reduced to about half of its original size. As an answer to this proposal, a coalition of NGOs prepared a so-called 'Shadow List', which is to be considered by the Polish government in the next stages of the creation of the Natura 2000 network. Because of their top–down character, however, many programmes of nature conservation turn out to be ineffective. Recent bottom–up initiatives of nature and landscape protection by individual regions and individual communes seem to be more promising and productive. This chapter focuses on one of the best examples, the UNESCO Biosphere Reserve West Polesie created in 2002 and situated in the poorest region of the EU.

In his contribution, Henny van der Windt offers a theoretical framework to assess and evaluate the Polish case. He also presents comparative findings from a Dutch case study – the Green River Plan – to place the Polish case in perspective. Van der Windt identifies three modes of decision-making, a top–down mode, a corporative mode, and a deliberative mode. These different modes vary with respect to the influence of the relevant actors in the decision-making process, the procedural style of governmental organizations, the (legal and financial) resources available to the different actors, and the range of issues that are taken into account. Van der Windt concludes that the decision-making process in both the Dutch and the Polish case can be characterized as a mixture of the top–down, the corporative and the deliberative mode. Interestingly, the Polish case differs from the Dutch case because of the relatively large role of science, a substantive rather than a procedural source of legitimacy.

Elizabeth Oughton and Jane Wheelock discuss the implementation of conservation policies on the North York Moors. They focus on the perceptions that a small group of stakeholders have of the conservation activities in the national park in which they live. This chapter looks at how policies are experienced by those on the ground. By taking a bottom–up approach, the authors are able to explore the net effect of all relevant policies and their outcomes. They examine views of legitimacy by exploring the perceived claims that people have on the landscape and environment – their entitlements. First, they look at the ways in which policy implementation affects such claims and the ability of local people to achieve their objectives with respect to the goods and services that the environment offers. Then, the authors explore the ways in which actions by locals can support or undermine the conservation objectives of international and national policy-making bodies. The chapter is based on a rural community in the North York Moors in England and uses as a specific case study the activities surrounding the conservation of heather moorland.

In her comments, Marielle van der Zouwen examines whether and to what extent Oughton and Wheelock's case study gives evidence of the two shifts that are at the heart of this volume, the shift from government to (multi-actor and multi-level) governance, on the one hand, and the shift from substantive to procedural sources of legitimacy, on the other. Van der Zouwen concludes that both shifts have only partly been realized. Consequently, present-day conservation practices are complex mixtures of traditional and new conservation policies.

1.3.3 Countries

This part opens with a chapter by Dirk Bogaert and Pieter Leroy, who look at the problematic application of Natura 2000 in Flanders. Due to a lack of public support, the implementation of its Flemish version, the Green Main Structure (GMS), failed.

As Europe expected Flanders to take its part in Natura 2000, the aim of an ecological network remained on the Flemish political agenda. Hence, the Flemish Ecological Network (FEN) that was launched after the failure of its precursor, was in fact a kind of replica of the former initiative. With lessons drawn from the policy disaster of the GMS, a uniform policy implementation and a clear communication plan were identified as crucial for acquiring legitimacy.

However, despite the lessons learned from the former failure, implementation of the FEN also run into severe difficulties. Because the policy design made no clear distinction between the issues of allocation, location, and joint use, it provoked protest and debate at all levels and on all issues simultaneously. In addition, political disagreement led to incompatible rules for public participation, increasing obstruction rather than reducing it. As a result, the handling of the FEN significantly damaged the legitimacy of Flemish nature policy.

In his comments on Bogaert and Leroy, Erik van Zadelhoff questions their diagnosis that attributes the failure of Flemish nature policy exclusively to flaws in the policy process. Van Zadelhoff compares the realization of an ecological network in Flanders and in the Netherlands and points to three specific differences that could explain why nature policy in Flanders was less successful than in the Netherlands. In the Netherlands, there is greater public support for nature conservation. Moreover, the Netherlands has a much stronger tradition in spatial planning, not only with respect to urban areas but also with respect to the countryside. Lastly, during the 1990s, concern for the environment was no longer the domain only of left-wing one-issue parties in the Netherlands, but had become a mainstream political topic.

Not unlike the situation in Flanders, in France the very first effort made by the government to implement Natura 2000 also met with fierce resistance. Initially, a sharp distinction was made between the compilation of national inventories by scientists and their validation at the European level, on the one hand, and the consultation involving other rural stakeholders in defining, drawing and developing management plans for these sites in a later stage, on the other hand. So the implementation of Natura 2000 in France at first favoured the scientific dimension at the expense of the social dimension, thereby provoking strong opposition by a powerful alliance of hunters, foresters, farmers and fishermen, that reached its peak in March 1996, when the administration presented its final list of sites covering 13 per cent of the national territory. Their protest led to a suspension of the inventory procedure for several months in 1997. To resume dialogue with the different stakeholders, this procedure was thoroughly revised. The consultation process, initially limited to management plans, was broadened to include the selection and size of sites. As in most other European countries, what took place was a shift from substantive sources of legitimacy, in particular scientific expertise, to procedural sources of legitimacy (to consultation, negotiation and deliberation), and, at the same time, a shift within the latter (procedural sources of legitimacy) from output legitimacy to input and throughput legitimacy.

In this chapter, Florence Pinton describes some consequences of this shift for the implementation of Natura 2000 in France. She presents a somewhat mixed picture. On the one hand the new system appears to be locally accepted, but the total area of sites proposed to the EU dropped from the initial figure of 13% in 1996 to 7% at the end of 2002. Despite a number of innovations, that have opened up space for new and promising forms of cooperation, mutual learning and consensus building, Pinton comes to the conclusion that the situation at this moment is far from ideal due to the very low involvement of scientists and experts and the fragmented representation of environmental and nature protection interests.

In his comments on Pinton's study, Henk van den Belt poses two important questions. First, he notices that the shift in emphasis from a top–down to a bottom–up approach everywhere in Europe seems to go hand in hand with a 'dilution' of the original nature goals, quantitatively (in terms of hectares) as well as qualitatively (from deeper to lighter shades of green). Van den Belt finds it rather unfortunate that Pinton refuses to take a normative stance on this issue. Secondly, Van den Belt questions the democratic quality of the consultation process with local stakeholders. He has strong reservations about the credibility of claims that the local knowledge of farmers, foresters and hunters is more adequate than 'abstract' and 'formal' scientific knowledge. Furthermore, the consultation process, Van den Belt contends, is about bargaining rather than arguing. It falls severely short of the criterion of inclusiveness because the fate of nature and biodiversity is actually placed in the hands of that 10% of the population who happen to live in the countryside.

In the final chapter of this part, Markus Leibenath discusses nature politics in Germany, a country with a decentralized, federal structure. Here, nature politics has turned into a complex multi-level policy arena in the past few years, or more accurately, into a complex set of interlinked horizontal and vertical policy arenas at different levels. Traditionally, nature conservation falls within the province of the Federal Länder. However, the Länder and the federal government also share some competences. With the introduction of a European biodiversity policy, a third level has been added. This new level has created a completely novel situation for the conservation authorities at the Länder level. While they used to be autonomous to a very large extent, the conservation authorities now have to coordinate the selection and delineation of protected areas with the Federal government and with European institutions.

Leibenath shows that in Germany too we can see a shift from substantive to procedural sources of legitimacy. He also makes it clear, however, that the EU has played no significant role at all in generating this shift. Implementation of Natura 2000 frustrated rather than facilitated such a shift and the accompanying improvement of input and throughput legitimacy. His case study on the implementation of Natura 2000 in the Federal Land of Brandenburg convincingly illustrates that the European Birds and Habitats Directives and the related case law leave little room for deliberation and consultation in the site selection process, as they were based upon a merely ecological-technical approach towards conservation, and prescribed a very tight time schedule. With the example of the establishment and management of the Müritz National Park, Leibenath demonstrates that more traditional instruments of German conservation policy can boast of a far better record in improving input and output legitimacy than Natura 2000.

According to Michiel Korthals, Leibenath's approach to conflicts over issues of nature policy suffers from a too narrow focus on distributive justice. To understand these conflicts, one should not only take into account the unequal distribution of power between stakeholders but also the uneven representation of visions of nature of the various stakeholders that determine what questions are considered relevant and what answers are seen as reasonable. As a source of conflict over nature policy, Korthals insists that misrepresentation is as important as misdistribution.

1.4 Conclusions

In the concluding chapter, Rüdiger Wurzel reviews the main insights developed in the preceding chapters while presenting an analysis of Natura 2000 within the wider EU environmental policy and political context. He traces the origins of Natura 2000 and its implementation measures to different phases of EU environmental policy. This chapter explains the main characteristics of and trends within the different phases of EU environmental policy (e.g. adolescent phase, mature phase and defensive/reorientation phase) while arguing that a focus on these phases helps us to gain a better understanding of the adoption, implementation and revision of Natura 2000.

This chapter also takes up the debate about the alleged shift from government to governance while explaining how this shift affected the implementation and revision of Natura 2000 measures within different phases of the multi-level EU environmental policy-making process. It explains the difficulties various member states have had in implementing Natura 2000 measures with the core features of the EU's multi-level decision-making process. For example, common implementation failures, which occurred in a wide range of countries or regions (as highlighted in several preceding chapters of this book), will be assessed with reference to the different and, at times, conflicting top–down and bottom–up strategies that were put forward by different EU or member state actors.

The conclusion reflects on the main themes of the preceding chapters, which include legitimacy, policy (or state) failure, trust, scientific versus local knowledge and bargaining versus deliberation. Finally, the conclusion highlights a number of important issues and research questions that were not (or only briefly) mentioned by the preceding chapters, for instance, the. North–South dimension and the EU as the optimal regulatory arena. In doing so, the conclusion asks whether there may have been incidents during the period covered by the high quality empirical research carried out by the authors when 'the dog did not bark'.

References

Alphandéry, P. and A. Fortier (2001). 'Can a Territorial Policy be Based on Science Alone? The System for Creating the Natura 2000 Network in France'. *Sociologia Ruralis* 41: 311–28.

Barber, B. (1984). *Strong Democracy. Participatory Politics for a New Age*. Berkeley CA: University of California Press.

Benz, A., F. Scharpf and R. Zintl (1992). *Horizontale Politikverflechtung : Zur Theorie Von Verhandlungssystemen*. Frankfurt/Main: Campus.

Benz, A. (1998). 'Politikverflechtung Ohne Politikverflechtungsfalle-Koordination Und Struk-turdynamik Im Europäischen Mehrebenensystem.' *Politische Vierteljahresschrift* 39(3): 558–89.

Benz, A. (2000). 'Two Types of Multi-Level Governance: Intergovernmental Relations in German and EU Regional Policy.' *Regional and Federal Studies* 10(3): 21–44.

Benz, A and B. Eberlein (2001). 'The Europeanization of Regional Policies. Patterns of Multi-Level Governance.' *Journal of European Public Policy* 6(2): 329–48.

Bohman, J. (1998). 'The Coming of Age of Deliberative Democracy.' *Journal of Political Philosophy* 6(4): 400–25.

Crouch, C. and W. Streeck (1993). *Industrial Relations and European State Traditions*. Oxford: Oxford University Press.

Elster, J., ed. (1998). *Deliberative Democracy, Cambridge Studies in the Theory of Democracy*. Cambridge: Cambridge University Press.

Friedrich, C.J. (1972). *Tradition and Authority*. London: MacMillan.

Glick Schiller, N. and A. Wimmer (2003). 'Methodological Nationalism, the Social Sciences, and the Study of Migration. An Essay in Historical Epistemology.' *International Migration Review* 37(3): 576–610.

Gouldner, A.W. (1970). *The Coming Crisis of Western Sociology*. London: Heineman.

Hampsher-Monk, I. (1992). *A History of Modern Political Thought. Major Political Thinkers from Hobbes to Marx*. Oxford: Blackwell.

Hooghe, L. and G. Marks (2001). *Multi-Level Governance and European Integration*. Lanham, MD: Rowman and Littlefield Publishers.

Kerr, A. (1960). *Industrialism and Industrial Man*. Cambridge MA: Harvard University Press.

Kersbergen, K. van and F. van Waarden (2001). *Shifts in Governance. Problems of Legitimacy and Accountability*. The Hague: NWO.

Kjær, A.M. (2004). *Governance*. Cambridge: Polity.

Marks, G. (2001). 'Types of Multi-Level Governance.' *European Integration online Papers* 5(11): 31.

OED. (1989). *Oxford English Dictionary*, digital version, second edition 1989, Oxford: Oxford University Press.

Oakeshott, M. (1962). *Rationalism in Politics, and Other Essays*. London: Metheun.

Ohmae, K. (1995). *The End of the Nation State: The Rise of Regional Economies*. New York: Free Press.

Osieck, E.R. (1998). Natura 2000: naar een Europees netwerk van beschermde gebieden. *De Levende Natuur* 99(6): 224–31.

Paavola, J. (2004). 'Protected Areas Governance and Justice: Theory and the European Union's Habitats Directive'. *Environmental Sciences* 1: 59–77.

Parsons, T. (1966). *Societies. Evolutionary and Comparative Perspectives*. Englewood Cliffs, NJ: Prentice Hall.

Pierre, J. and B. Guy Peters (2000). *Governance, Politics and the State*. London: MacMillan.

Pierre, J., ed. (2000). *Debating Governance. Authority, Steering, and Democracy*. Oxford: Oxford University Press.

Pous, P. de and A. Beckmann (2005). *Natura 2000 in the New EU Member States*. Brussels: WWF.

Putnam, R. (1988). 'Diplomacy and Domestic Politics : The Logic of Two-Level Games.' *International Organization* 42(3): 427–60.

Rawls, J. (1995). *Political Liberalism*. Harvard: Harvard University Press.

Rhodes, R. (1997). *Understanding Governance: Policy Networks, Governance, Reflexivity and Accountability*. Milton Keynes: Open University Press.

Rueschemeyer, D. and T. Skocpol (1996). *States, Social Knowledge and the Origins of Modern Social Policies*. Princeton: Princeton University Press.

Scharpf, F. (1970). *Demokratietheorie Zwischen Utopie Und Anpassung*. Konstanz: Universitätsverlag.

Scharpf, F. (1988). 'The Joint-Decision Trap.' *Public Administration Review* 66(3): 239–78.

Scharpf, F. (1997). *Games Real Actors Play. Actor-Centered Institutionalism in Policy Research.* New York: Westview Press.

Scharpf, F. (1999). *Governing in Europe. Effective and Democratic?* Oxford: Oxford University Press.

Scharpf, F. (2003). 'Problem-Solving Effectiveness and Democratic Accountability in the EU.' *MPIfG Working Paper* 35. Cologne.

Scharpf, F. (2004). 'Legitimationskonzepte Jenseits des Nationalstaats.' *MPIfG Working Paper* 35. Cologne.

Schmitter, P. (1974). 'Still the Century of Corporatism?' *Review of Politics* 36: 85–131.

Schmitter, P. (2000). *How to Democratize the European Union... And Why Bother?* Boston: Rowman and Littlefield.

Scott, J. (1998). *Seeing Like a State. How Certain Schemes to Improve the Human Condition Have Failed.* New Haven/London: Yale University Press.

Smith, A. (1983). 'Nationalism and classical social theory.' *British Journal of Sociology* 34:19–38.

Strange, S. (1996). *The Retreat of the State. The Diffusion of Power in the World Economy.* Cambridge: Cambridge University Press.

Streeck, W. and P. Schmitter, ed. (1985). *Private Interest Government. Beyond Market and State.* London: Sage.

Weber, M. (1972). *Wirtschaft und Gesellschaft; Grundriss der Verstehenden Soziologie.* Tubingen: J.C.B. Mohr (originally published in 1921).

Part II
Protected Species

Wintering Geese (by Melchert Meijer zu Schlochtern)

Chapter 2
Wintering Geese in the Netherlands

Legitimate Policy?

Gilbert Leistra, Jozef Keulartz and Ewald Engelen

2.1 Introduction

In November 2003, the Dutch Minister of Agriculture presented the next phase in almost half a century of 'handling' the wild geese populations in the Netherlands in the Policy Framework Fauna Management (PFFM). Whereas in the 1970s geese were put on the (inter)national policy agenda due to the alarming decline in wild populations as a result of excessive hunting, the cause of concern at the beginning of the twenty-first century is not their threatened survival but their success at it. The international and national initiatives taken from the seventies onwards have proven to be so successful that Dutch wild geese populations have increased by a factor of ten, which has dramatically augmented damage to local agricultural crops. The mounting costs of indemnification and the growing distress of the agricultural community confronted with wintering geese populations led to the PFFM, which establishes foraging areas for these geese. The PFFM focuses on (i) providing the agricultural community with a durable solution while at the same time, (ii) safeguarding the geese populations, and (iii) keeping the costs of indemnification in check.

The establishment of foraging areas for wintering geese can be seen as representative of the many nature conservation policy processes in the Netherlands. It is characterized by (inter)national goals and responsibilities, processes at different administrative levels and a multitude of actors with their own specific interests. The many coordination and distribution issues that surface throughout the policy process at all administrative levels raise strong legitimacy requirements! The case of wintering geese provides an excellent opportunity, we contend, to gain insight into the problems and possibilities of legitimacy production in what is clearly a multi-level, multi-agent and ultimately, a highly controversial, arena of policy making and policy implementation.

The structure of this chapter is straightforward. We start with a section that places the theoretical notions of legitimacy, as discussed in detail in chapter one, in relation

Gilbert Leistra
Applied Philosophy Group, Wageningen University
E-mail: gilbert.Leistra@wur.nl

J. Keulartz and G. Leistra (eds.), *Legitimacy in European Nature Conservation Policy:*
Case Studies in Multilevel Governance, 25–45. © Springer 2008

to the character of nature conservation in the Netherlands. The next section forms the core of the chapter and describes in some detail the implementation of the PFFM. In the final section the legitimacy of the PFFM is assessed. Rather than giving a simplistic normative judgment, the assessment tries to highlight the problems and possibilities of legitimacy production in nature conservation policy.

2.2 The Issue of Legitimacy in Dutch Nature Policy

As outlined in chapter one, legitimacy refers to the question of why the outcomes of binding collective decision-making ought to be accepted by those whose interests are being harmed by the decision in question (Scharpf 1970; 1999; 2004; Weber 1972: 16–17, Ch. 9). In modern-day Western democracies, individual interests are seen as legitimate concerns, which cannot be harmed without valid reasons. Consequently, when there is a higher order commonality (society) whose interests could collide with those of the constituting individuals, well-founded arguments need to be provided to legitimize outcomes of collective decision-making that place the collective concern over and above individual interests.

Acceptance can be regarded as a measure of legitimacy. When a policy is accepted without objection, or with just minor objections even by those that are negatively affected, one can assume that well-founded arguments were convincingly provided that placed the collective concern over and above the concerns of the affected parties. However, the absence of acceptance and the resistance to decisions or policy of democratic authorities does not automatically imply an absence of legitimacy and subsequently the absence of the 'right' arguments for pursuing a specific course of action as set out in a decisions or policy. This issue of legitimacy especially arises in 'dilemma'- and 'zero sum'-games (Scharpf 2004). In the former, the binding force of collective courses of action is severely weakened by the availability of opportunistic choice options, and in the latter, a redistribution of resources from some agents to others is implied. Unpopular policy and decisions are not unthinkable in these situations, even if they provide an often necessary and urgent remedy to societal problems. Nonetheless, in today's democratic society, the state is no longer omnipotent and omniscient, it has lost its monopoly on policy making to other public and private agents and has increasingly become embedded in a multi-level governance structure that combines, in a complex and incoherent way, territorial and functional modes of organization.

The above implies that well-founded arguments need to be provided for both the outcomes/effects of the intended decision and the procedures that led up to the decision. A government cannot base its legitimacy on output or efficiency alone if it wants to call itself a democracy. Democratic regimes gain their legitimacy from different types of arrangements, through majority rule and the protection of minorities as well as through procedures and securing the needs of society at large. Legitimacy can, therefore, be seen as a function of the interdependence and interplay of procedures (input and throughput) and results (output) (see chapter one for more detail). The assumption, as clarified in chapter one, is that whether one likes it or not legitimacy increasingly has to be produced procedurally since substantive sources of legitimacy – religion, charisma, tradition, scientific expertise (Weber 1972: Ch. 9; Friedrich 1972) – which often establish the foundation of many policy and political processes and outcomes, have increasingly lost their legitimating force. The limits of rationality and the multitude of actors at different administrative levels increase the need for more emphasis on the input and throughput requirements of legitimacy production. This need can also be observed in Dutch nature conservation policy.

The birth of what could be called the Dutch tradition of nature conservation policy dates from the late nineteenth century. In line with the liberal hegemony of those days and its shibboleths of the night watchman's state, Dutch environmentalism was initially a purely private matter; enlightened citizens, belonging to the well-to-do layers of Dutch society, started to value the romantic characteristics of some parts of the Dutch landscape and aimed to protect it from the effects of industrialisation and urbanisation. While the introduction of universal voting rights in 1918 slowly changed the perspective on what was and was not a legitimate task for the state, it took until after the German occupation in the Second World War before environmental protection and the management and conservation of natural resources was institutionalized as an official policy objective. However, it was not until the late 1980s that Dutch nature policy matured. The 1990 Nature Policy Plan (NPP), which embodied the maturation of Dutch nature policy, was epochal in more than one respect. Characteristic of the 1990 plan is an ecological ideal that was used as a benchmark: a scientific reconstruction of what living nature under given physical conditions would have looked like in the absence of human influences. This ecological norm was presented as an objective criterion. Considerations of cultural history did not matter; nature was not a matter of taste. In order to give nature development a chance, the plan claimed a considerable amount of space in the form of the National Ecological Network (NEN). This network, composed of 'core areas', 'nature development areas' and 'ecological corridors', proved to be the exemplar for Nature 2000.

As such, the 1990 plan is a textbook example of top–down planning, based on substantive sources of legitimacy. On the assumption that it is up to scientific experts and not ordinary citizens and politicians to determine the direction of nature policy, the premise underlying the entire NPP process was the superiority of expert ecological knowledge about the various ecosystems and the environmental conditions in which they are viable over other types of claims. However, the scientific considerations proved insufficient to trump individual interests. As was more generally the case in Dutch planning in the 1990s (spatial as well as environmental planning), the implementation process of the 1990 Nature Policy Plan came to a virtual standstill as local stakeholders increasingly used their powers of obstruction to protect their interests.

In response, the government gradually abandoned its centralist, top–down planning approach and increasingly switched to methods of participatory and interactive policy-making (Hajer and Wagenaar 2003; Akkerman et al. 2004). This change reflected, in the terms used in chapter one, a replacement of substantive sources of legitimate policy making with procedural ones. The first result of this shift was the 2000 report 'Nature for People, People for Nature' from the Dutch Ministry of Agriculture. Key passages in the report stressed that 'Nature should be at the heart of society' and that 'Nature should be more strongly anchored in the hearts of the people and in the decisions of citizens, entrepreneurs, social organizations, and local authorities' (LNV 2000:18). To achieve this goal, the government opted for a substantial 'broadening' of the aims of its nature policy. On the one hand, 'creating' nature was presented as being not only and not primarily a good in itself, but as something with a clear instrumental value; nature should primarily be for the benefit of Dutch citizens. This is the 'nature for people' part of the title, and, as such, went against the grain of the main tenets of the ecologists and environmentalists who had long dominated Dutch nature policy. However, more traditionally, it also stressed the conservationist responsibilities of the Dutch state and its citizens for a sustainable use of its natural resources; the 'people for nature' part of the title.

In recent decades, we have seen a clear transformation of Dutch nature conservation policy from an explicit top–down approach, which derived its legitimacy from substantive sources, in particular scientific expertise, to a much more procedurally oriented policy-making mode in which individual interests, even those that ran counter to ecological and environmental considerations, were accepted as legitimate claims and hence would have to be included in the policy-making process. In this chapter we will see whether, and by what means, this shift in policy strategy has taken effect in current nature conservation policy practices by looking at the exemplary case of foraging areas for wintering geese. When legitimacy is seen as a function of the interdependence and interplay of procedures (input and throughput) and results (output), then an assessment of the legitimacy of a policy should address both procedures and results. This case provides an excellent opportunity, so we contend, to gain insight into the problems and possibilities of legitimacy production in what is clearly a multi-level, multi-agent and ultimately, a highly controversial arena of policy making and policy implementation.

2.3 Implementing the Policy Framework Fauna Management

2.3.1 Wintering Geese

For centuries, geese have been attracted to the Dutch delta. The delta's geographical location has always been an important node in the transnational networks of migratory birds. It provides safe sleeping areas on the larger lakes and sandbars, and plenty of food, which is available on agricultural lands and salt marshes. Currently, 1.5 million wild geese, half of all wild geese wintering on the European continent, winter in the Netherlands (Sanders et al. 2004), which emphasizes the responsibility of the Netherlands in protecting wild geese populations.

The flourishing geese populations that the Netherlands is currently harbouring are only partly due to its geographical location. In the late 1960s, the geese populations started to show clear signs of decline as a consequence of excessive hunting. In a context of increasing environmental awareness among Dutch policy makers and environmentalists, the Dutch government in the late 1970s took the initiative to launch a supra-national agreement at the EU-level to provide migratory birds with a larger number of safe havens that could help sustain their numbers. This initiative was highly successful. The EU Birds Directive of 1979 was one of its immediate outcomes, but the Habitats Directive of 1992 and the African Eurasian Migratory Waterbird Agreement of 1999 can also be seen as building upon this first legislative activity. The Netherlands was thus one of the first countries where the decline in wild geese populations was perceived and translated into political action. The protective regime, which was established in the Netherlands and surrounding countries from the 1970s onwards, resulted in a magnificent recovery of the wintering geese in Western Europe, where most of the once decimated populations increased by a factor of ten. Emblematic of this success is the Graylag goose (*Anser anser*). With an average wintering population size of 10,000 in the 1960s, 100,000 in 1990, the population grew to 244,000 geese in 2002 and shows no sign of decline in growth (Koffijberg et al. 1997; Van Roomen et al. 2004), which makes the population's recovery undeniable.

While the protective regime resulted in a recovery of the geese populations, it also led to increasing problems for the agricultural community, which had to deal with the growing populations that foraged on their lands. In 1975 measures were taken to tackle the problem of increasing damage to agricultural crops by establishing a reserve on Texel, one of the Dutch '*wadden*' islands, that allowed Brent geese to forage freely without disturbance. This so-called 'stick and carrot' method allowed the farmers on the island to chase geese off their lands (the 'stick'), to the reserve which was specially managed for the then vulnerable Brent geese (the 'carrot'). Ever since the 1970s the agricultural damage inflicted by geese has been steadily increasing. This is due to the expanding population of geese, but also finds its cause in the intensification of the cattle farming industry, which has led to an increased demand on the existing grasslands (Groot Bruinderink 1987). The 1990 policy document 'Space for Geese' (LNV 1990) tackled this emergent problem and sketched out how the Netherlands has been dealing with wild populations of wintering geese in the last decades by (i) emphasizing the international responsibility in protecting wild geese, (ii) indemnifying all damages to agricultural practices, and (iii) encouraging regional cooperation in the protection of the more vulnerable populations, which lead to the establishment of several reserves including the one on Texel. However, in the nineties the geese populations increased in size even more and consequently so did the costs of indemnification. Although all costs were compensated, the frustration of the agricultural community remained as they had to increase their efforts in dispelling geese before being eligible for remuneration (Ebbinge et al. 2003).

The successes of the conservation policies in the Netherlands started to attract increasing attention from Dutch MPs, who feared increasing costs. In the period 1990–2002, a total of €35 million was spent as a result of wildlife damage. Of this damage, 85% was credited to waterbirds, with geese as the major wrongdoer, causing €23 million (65%) in damage. The species Greylag Goose (*Anser anser*), White-fronted Goose (*Anser albifrons*) and the Eurasian Wigeon (*Anas Penelope*) were responsible for 53% of all damage (Ebbinge 2003).[1] In 2002 the annual total for these three species amounted to a little over €2 million, €500, 000 more than in 1996 and a stunning €1.5 million higher than in 1990, while no end to its rise appeared imminent (Ebbinge et al. 2003).

This situation climaxed in 2003 when a number of MPs from the parties in office filed a motion that left nothing open to interpretation, calling for a reopening of the hunt on the three most harmful species in order to contain future costs. The aforementioned motion, which proposed a reopening of the hunt on these three species, was at the same time a motion of distrust regarding the 12 provinces in dealing with wildlife damages adequately. The Flora and Fauna Act (FF-Act) of 2002 adjudges the provincial authorities responsible for dealing with wildlife damage. This responsibility is in line with the Decentralization Agreement of 1995, which was established between the national government and the Interprovincial Consultative Body (organization representing the twelve Dutch provinces) and which aims at policy that is attuned to regional characteristics and problems. The controversy concerning the distribution of responsibilities in November of 2002 is remarkable since the motion was filed and adopted just months after the FF-Act took effect. Although the hunting of these species is not prohibited under the Birds Directive, the Dutch Minister of Agriculture deemed a reopening of the hunt as inconsistent with the character of the international responsibility of the Netherlands and argued that the 12 provinces should be given enough time to implement the FF-Act correctly. However, the concern of increasing costs proved to be real when in 2003 the total amount of damage caused by above species amounted to a total of €4 million (Faunafonds 2004). This was twice the amount of the previous year and the Minister was urged to take action once again.

Despite Dutch vanguardism in the field of supranational environmental policy making and the success story of the wintering geese, the implementation of supranational agreements within its own territorial and jurisdictional domain has been much less energetic (Zouwen and Tatenhove 2002). The EU directive relevant here, the Birds Directive, was only converted into Dutch law in 2002.[2] In that year, the Dutch Flora and Fauna Act (FF-Act) was enacted. This provided a legal framework for

---

[1] From now on, the species Graylag goose, White-fronted Goose and the Eurasian widgeon will be referred to as 'geese'. Although the widgeon is not a goose, the PFFM is also focused on reducing the cost widgeons account for. Within the policy documents and related correspondence references, the term 'wintering geese' refers to these three species.

[2] Together with the Habitats Directive, the Birds Directive shapes Natura 2000. Natura 2000 distinguishes between two regimes of protection; (i) the protection of wild species living within the European Union and (ii) the protection of habitats. These international obligations have been translated into national legislation; the Flora and Fauna Act, which took effect in April 2002, deals with the protection of wild species, whereas the amended Nature Protection Act, which took effect on the first of October 2005, provides the legal framework for the protection of habitats. With the implementation of these acts, adequate national legislation should exist to offer enough guarantees for the protection of European nature in the Netherlands.

the distribution of responsibilities between the different layers of the Dutch polity, i.e. (central) state, provinces and local municipalities. As such, the case of wintering geese does and does not illustrate the increasing multi-layeredness of many policy domains because of European integration. On the one hand, the Dutch legal initiatives clearly possess an international dimension. On the other hand, EU directives, in particular the Birds Directive, are conceived in such a fashion as to accommodate national sovereignty over internal affairs, demonstrating that strong claims about the demise of the nation-state are overstated (Strange 1996).

2.3.2 Policy Framework Fauna Management

The pressure on the Minister of Agriculture to take action ultimately resulted in the Policy Framework Fauna Management (PFFM) in November 2003, which introduced the next chapter in geese policy in the Netherlands. The consultation process that led up to the PFFM involved a large number of public and private agencies, ranging from farmers' representatives to environmental associations. The PFFM stressed the overall responsibility of the central government, but at the same time committed the other parties involved to enact locally what was being decided centrally and endorsed them to take part in the committee that was to aid and advise the Minister during the implementation process. Concerning the issue of rising costs, it was decided that the provinces would demarcate 80,000 hectares, most of which is farmland, as special foraging areas for wintering geese. Only within the boundaries of these foraging areas will farmers have the right to indemnification for damages and other costs. As such, a farmer located in an established foraging area can expect to be compensated for crop damage and will receive a management fee and an additional stimulant bonus for participating. Although the exact amounts have not yet been formally agreed upon and vary between arable farming lands and pasture lands, a farmer can expect a total of about €250 per ha a year when participating in the PFFM (Ebbinge 2004). Outside these areas, farmers will be allowed to dispel wintering geese and to shoot them if necessary to prevent crop damage, but they can no longer rely on indemnification, even if their efforts prove to be fruitless. By limiting the surface area that falls under the indemnification plan, the Ministry of Agriculture hoped to provide wintering geese with enough food and space to fulfil the Dutch government's international obligations, while at the same time restricting its seemingly endless financial obligations. While formerly the whole of the Netherlands was, at least in principle, a foraging area for migratory geese, under the PFFM only a small part of the Netherlands will provide a safe haven for these birds. This strategy assumes that this large scale 'stick and carrot' strategy does provide sufficient surface area and succeeds in binding the wintering geese to these intended foraging areas.

Although the PFFM builds on decades of experience with protecting wild geese populations and the damages these populations inflict, the solutions presented radically increased the legitimacy requirements for the policy in question. Noteworthy about this policy shift is that, in terms of Scharpf's analytical distinction presented earlier in chapter one, the new policy represented a shift from a pure coordination game without distributive consequences to a zero-sum game with evident distributive effects. Under the earlier policy, all farmers stood to gain while the taxpayer was footing the bill. In other words, there were real and immediate gains for those directly involved and only small and diffuse costs that were being distributed among a large number of anonymous taxpayers. The effect was a low level of politicization and hence very lax legitimacy requirements. Under the new policy, on the other hand, some farmers stand to gain while others stand to lose. As such, there have to be good reasons for the losses of the one and the gains of the other in order to overcome the obstructive power of the losers. In the absence of good substantive reasons – the equivalent of our substantial sources of legitimacy – the alternative to legitimize such a policy would be the production of legitimacy by procedural means. We will now show how the Dutch government tackled these increased legitimacy requirements.

Initially, the PFFM included three phases of implementation with rather strict time-paths. During the first phase, the provinces were given the task of creating the legal possibility of dispelling and if necessary hunting foraging geese, while locating the most suitable areas for demarcation. Compensation for damages would be paid only if sufficient precautionary measures had been taken. This would take until October 2004. During the second phase, the provinces were supposed to appoint enough foraging areas to fulfil the quota given to them by the Ministry of Agriculture. This was to be finished by June 2005. Subsequently, in order to fulfil formal requirements, the decision to appoint certain locations as foraging areas had to be formalized in 'Nature Area Plans'. From this time on the farmers within a foraging area could apply to Subsidy Schemes Agricultural Nature Management (SAN) as a collective. By formally anchoring the foraging areas in the provincial Nature Area Plans and developing the SAN subsidy packages, part of the costs involved in the management of foraging areas are now being passed on to the European Union. SAN packages are co-financed by the European Union through the European Agricultural Fund for Rural Development (EAFRD). Whereas the costs were predominantly a burden of the Dutch taxpayer, this new strategy entails a shared financial responsibility in protecting 'European nature' in the Netherlands. To be eligible for this co-financing, the packages need to be approved by the European Commission. The European Commission will test the subsidy schemes to see whether these do or do not entail unfair income support to Dutch farmers. However, upon presentation of the PFFM, the SAN packages were not yet fully developed and consequently have not yet been approved by Brussels, resulting in uncertainty regarding the amount of the actual subsidies.

In the autumn of 2004, it still looked as if everything was on schedule. According to a report by the Ministry of Agriculture to the Dutch parliament, nine of the twelve Dutch provinces would succeed in appointing their quota of foraging areas before the end of the year. However, it soon appeared that the Ministry had been too optimistic. A number of MPs became aware of huge resistance to the top–down approach by the farmers as well as by the provincial authorities. The resistance was fuelled by the numerous uncertainties that the PFFM contained upon presentation.

The uncertainty regarding the actual SAN packages, disagreement about the actual amount of the subsidies, the absence of approval from Brussels and the fear that participation in the PFFM would lead to permanent restrictions to their agricultural practices, caused farmers throughout the Netherlands to protest against the appointment of their enterprises as foraging areas. The fact that many farmers saw their businesses being established as foraging areas on planning documents, without having been consulted, increased their resistance even more. The provincial authorities, responsible for the actual demarcation in their province, were only equipped with an incomplete story and therefore incapable of quelling the fears of the local farming communities. As a result, the authorities increasingly turned to the Ministry of Agriculture to take over responsibility.

Needless to say, clear signs were already given within the first phase of the implementation of the PFFM by both the agrarian community and the twelve provinces that the implementation process was going to be problematic. In response, the Dutch House of Representatives stressed the need to secure local support, implying a replacement of the traditional Dutch top–down approach by a more collaborative approach in which local agents were invited to contribute to a collective solution that could meet a greater level of local commitment than could authoritative solutions. In the terms used in this paper, this change in policy represented a shift from more substantive sources of legitimacy production to procedural ones, a shift that was forced upon the Ministry of Agriculture by the fact that through its own policies, as we discussed above, it had changed the political setting from a coordination game into a zero-sum game.

The new focus on public support did not alter the initial timeframe. When in November 2004 the need for local support became explicit, the deadline of June 2005 was maintained. The Minister of Agriculture emphasized the importance of meeting this deadline as it would be followed by the official SAN application procedure. If this deadline was not met, farmers who were willing to participate in the PFFM would have to wait another year before they would be able to apply for SAN packages, which meant another year without co-financing with EU money. The Minster also stressed the necessity of the success of a swift implementation of the PFFM, as the costs of compensation had increased to €7 million in 2004. However, in 2004, the SAN packages still had not been approved by Brussels. This was problematic, as the success of the PFFM, not least the affordability of this framework, depends on the approval of Brussels. No approval. . .no co-financing! This has had the effect that the costs of the PFFM for the next annual cycle, fixed at €14 million a year, will be hard to justify. The estimated €14 million is a considerable amount more than any of the yearly indemnified sums, which is surprising since the ever increasing costs of indemnification were the mainspring of the strategy set out in the PFFM. The justification of this sum must be found in the assumption that with the PFFM in place (i) the agricultural community is offered a durable solution, (ii) the population of wintering geese is safeguarded and that (iii) the now fixed costs are being co-financed by the European community. This, however, remains to be seen!

In the following section, the process of implementation is discussed in three of the twelve provinces. Due to ecological and agricultural differences, the attractiveness of different parts of the Netherlands to the wintering geese populations varies. The PFFM takes this into account, as the distribution of the 80,000 hectares is based on so called 'geese days' (the number of geese multiplied by the number of days the geese are present in an area). Most provinces have been allotted a quota of between a 1,000 and 10,000 hectares. These quotas are, in fact, a good indication of the respective contribution of each province to the convergence of the problem at the national level, but also give an indication of the perceived problem at the provincial level. It should be obvious that the problem in Limburg, which has been allotted a quota of 600 hectares, is of a totally different order than that of the province of Friesland, which has been assigned the largest quota with 30,000 hectares. Limburg, Friesland, and Brabant (see Fig. 2.1) are chosen to highlight the regional differences and the consequences this has for implementing a uniform national policy in heterogeneous regional situations. The dissimilarities in the allocation of the 'geese' problem, and the history of dealing with this situation in different contexts, prove to be important in the success of the implementation of the PFFM.

2.3.3 Limburg

Where most provinces were struggling to assign their quota, Limburg proves to be the only province where the initial top–down approach was successful and it

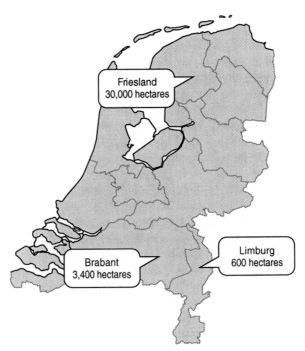

Fig. 2.1 Three provinces under investigation

managed to comply with relatively little effort and resistance to all requirements of the PFFM. This success needs to be understood in relation to the perceived problem of wintering geese in Limburg.

This most southern province of the Netherlands is the least attractive to wintering geese. The 600 hectares allocated to Limburg are by far the smallest quota and are less than 1% of the total 80,000 hectares needed for the wintering geese. Comparing the 600 hectares with the 30,000 hectares allocated to Friesland, it is evident that the perceived problems at the national level are not due to the size of the wintering populations and remunerated damages in Limburg. When confronted with the PFFM, Limburg immediately complied by assigning foraging areas on the basis of inventory data. No prior consultation took place, and once the locations were delineated, the planned areas were open to public review according to the Dutch Administration Act. A number of negative reactions forced the Limburg authorities to convene an information meeting where farmers were introduced to the PFFM and were given the opportunity to react to its content. This meeting was followed by negotiations between the provincial representatives, Birdlife, the Netherlands and representatives of the agricultural community, which resulted in an amendment of the initial demarcation of the areas. As soon as the foraging areas were delineated, the areas were incorporated into the Nature Area Plans of the province in accordance with the second stage of the PFFM. As the provincial authorities were able to comply with this timeframe, Limburg could now wait until the approval of the SAN packages by Brussels and start contracting the participating agrarian communities for periods of six years.

Judging from the relatively low resistance in Limburg to the top–down approach and ease of implementation, the level of politicization is apparently not high enough for the fairness of the policy to be contested. The relatively little resistance to the provincial top–down approach could easily be amended by a 'procedural' effort. The decision of who is to 'gain' – those who participate and receive subsidy – and who is to 'lose' – those who do not participate and have to make an effort in chasing geese away but receive no subsidy – was largely made by the provincial authorities. As indicated, this strategy proved to be unsuccessful in other provinces and led to a change of strategy: delineation on basis of support.

2.3.4 Friesland

Friesland is by far the most important province in the Netherlands for geese and has decades of experience in handling the growing geese populations. With a quota of 30,000 hectares, 37% of the total land dedicated to the wintering geese, it should be clear that the perceived problem is most urgent in this province. Whereas in Limburg geese are just one of the wildlife species that cause damage to agricultural crops, geese are the major culprit in Friesland and have been so for the last couple of decades. Under the policy of the 1990s, Friesian farmers were constantly occupied with scaring away geese in order to qualify for indemnification. With an ever increasing geese population and the almost exclusively agricultural character of the Friesian landscape, this meant driving geese in effect on to their colleagues' and neighbours' land. This was not considered to be a solution but just a continuous relocation of the problem. Moreover, the situation gave rise to heated disagreements about the total amount of compensation between the affected farmers and those responsible for the taxation of the damages. This ultimately led to a situation where the agricultural community was developing a negative attitude towards the geese, which worsened the relationship between the agricultural community, conservation agencies and the general population (Ebbinge et al. 2003).

On the initiative of several farmers, a breakthrough was made in 1996 when opportunities were explored with the Fauna Fund (the organization responsible for compensating wildlife damage) to initiate experiments with a focus on harbouring wintering geese. These experiments were, in fact, very similar to the current PFFM and entailed clear guidelines of how much compensation a farmer could expect if he would tolerate the geese on his land. This approach would not only generate clarity about the amount of compensation, but geese would also be left to graze freely on the lands of participating farmers without being scared away and thus without relocating the problem. The main difference between the 1996 policy and the PFFM is that in 1996, the farmers outside the protected areas, still suffering damages, were compensated. Under the PFFM, this will not be the case. Although the above arrangements did not remedy all problems and frustrations, Friesland is now able to look back on 10 years of positive experience with conservation measures. In fact, geese associations have been established within the agricultural communities and are new and cooperating partners in dealing with the wintering populations. As a result, the relationship between the agricultural community and the authorities in Friesland has improved (Ebbinge et al. 2003).

Besides the positive experiences with the geese arrangements, Friesland also had negative experiences regarding the top–down implementation of the NEN (National Ecological Network) in the mid-nineties, which required the establishment of 550 hectares of nature reserve on agricultural grounds in Gaasterland. The top–down plans infuriated the agricultural community and official maps delineating the reserve had the effect of a red rag to a bull. The resistance was enormous and stretched beyond the agricultural community to the general public, when it finally reached a deadlock. Only after the intervention of a group of 'wise men' were negotiations opened, which ultimately led to an abandonment of the initial plans of expropriation and long-term management by conservation agencies. Instead, the agricultural community was made responsible for the realization of ecological goals by means of short-term contracts (Kuindersma and Kolkman 2005). Although the ecological successes are contested (cf. Joustra 2001; Molenaar et al. 2005; Molenaar et al. 2006; Selnes et al. 2006), the Friesian authorities now use the Gaasterland experience in implementing many rural policies.

Equipped with the lesson learned that 'substantive' scientific arguments are not enough to legitimize and guarantee efficacious policy, the provincial authorities in Friesland were able to choose a strategy that was quite different from the other Dutch provinces. Almost immediately after the Ministry of Agriculture presented its plans, a top–down approach was deemed unacceptable and an alternative 'procedural' approach was taken. The chairmen of the associations – the new and co-operating partners – were invited for an informative meeting. Instead of presenting plans for implementation, the chairmen were asked to reflect upon the requirements of the PFFM and look for possibilities of implementation. The chairmen declined the challenge of establishing the foraging areas in their community, as they did not want to be held responsible within their own community for a government policy that entailed so many uncertainties. The uncertainties of the future consequences and possibilities of their agricultural practices were a cause for concern, but also the sizes of the subsidy packages were still indeterminate. As a result, the provincial authorities slowed down the process as they realized that without a clear message and without public support, the PFFM had no chance of success in Friesland.

When public support became an issue on the national agenda in the autumn of 2004, the provincial authorities were ready to take the next step. As the provincial organization was not equipped to arrange a large-scale process of delineation based on public support, BoerenNatuur (Farmer and Nature), an organization representing the interests of the agricultural and private nature conservation organizations, was contracted. This process was preceded by a clear message from the provincial authorities: foraging areas would not be assigned without support from the community and participation would not bring about future restrictions to agricultural practices. BoerenNatuur arranged for several meetings in each potential area, focused on information and education, to gauge ultimately the support in each area. In addition to the presence of wintering geese, the potential of each location was assessed on the basis of the presence, or possible establishment, of agrarian and public nature management organizations like the geese associations. In this respect, Friesland profited once more from the already established network of actors involved in geese management and the positive experiences from the past ten years. The next step was the delineation of a proposed foraging area by each community, accompanied by a request from that specific community to participate in the PFFM.

This process proved to be very successful and the 30,000 hectares allotted to Friesland were exceeded by 20,000 hectares for a sum total of 50,000 hectares. Where most provinces were struggling to get their quota assigned, Friesland was now faced with the task of reducing its 50,000 hectares down to 30,000 hectares. Friesland again contacted the geese associations, this time to discuss reduction. A short feedback process ultimately led to a delineation of 30,000 hectares. Only in the tougher decisions where the communities were not able to downsize the areas themselves did the provincial authorities step in and base their decision on 'substantive' arguments, using inventory data and damage figures.

We argue that the PFFM represented a shift from a pure coordination game without distributive consequences to a zero-sum game with evident distributive effects where farmers within the foraging areas stand to gain and those outside the foraging areas stand to lose. After all, participating in the PFFM is supposed to provide a farmer with a durable solution, which entails clarity about the amount of indemnification. The farmer outside the established foraging area, although able to dispel visiting geese, gets no indemnification when suffering damages. It appears that the participatory process in Friesland succeeded not only in finding support for 30,000 hectares but also in making clear that when a farmer is located within these 30,000 hectares, he stands to gain.

2.3.5 Brabant

Brabant has been assigned a quota of almost 3,400 hectares, which is considerably more than Limburg, but just a tenth of the amount that Friesland has been assigned. Although the frustrations of the agricultural community in Brabant are similar to the frustrations of their colleagues in Friesland, the provincial authorities in Brabant do not perceive the distribution problem as their colleagues at the national level do. This is illustrated by the reluctance of the Brabant authorities to adopt the PFFM. For example, in 2004 the Brabant authorities chose not to adopt the first phase of the PFFM and deemed a reopening of the hunt while searching for suitable foraging areas not in keeping with international obligations. With this choice they, in fact, took responsibility for the task that was assigned to them under the FF-Act and offered to indemnify farmers during the first phase of implementation while locating suitable foraging areas. The compensated losses amounted to €50, 000 that year and illustrate the relatively small contribution to the politicization of the problem at the national level. In fact, the need for a 'distributive' solution is not (yet) perceived in Brabant, but Brabant is forced under the PFFM to make their contribution to this solution by assigning 3,400 hectares.

Similar to the approach of the Limburg authorities, the foraging areas in Brabant were selected on the basis of ornithological inventory data and damage figures. However, it soon became apparent that scientific arguments were insufficient to convince farmers and their collective organizations to participate in the implementation. All 120 reactions to the intended plan were negative and a 'Limburg-like' quick fix seemed not only unwanted but also unrealistic. The resistance to the plan was and still is enormous, especially in the west of Brabant. In this part of the Netherlands, in contrast to most other areas, geese cause the most damage due to the arable land-use. Farmers on arable land not only experience income loss due to crop damage, but also need to take active action to restore the damage and ensure a continuing crop rotation. Moreover, the damages to arable land are less predictable, depict a more erratic character and are, in general, more substantial than damages to pasture lands. A spokesperson for the agricultural organization representing farmers' interests, the same organization that subscribed to the PFFM upon presentation, alleged that the provincial authorities made a mistake by delineating foraging areas in the west of Brabant. His message was that in the west of Brabant they would keep resisting the PFFM. Both the subsidy packages for arable land and the perceived sole responsibility of the agricultural community in dealing with wintering geese are contested. The spokesperson argued that the burden should not only be borne by the farming community but ought to be distributed equally among all large property holders, such as the state-owned Forestry Corporation and the private non-profit Conservation Foundation. In October 2004, the provincial authorities in Brabant decided to postpone the process of delineating foraging areas, and together with the provincial authorities in Friesland, they set the precedent for a motion of parliament that resulted in a strategy of implementation on the basis of support.

In view of this resistance, the provincial authorities nullified their initial plans and set up a deliberative consultation process instead. The professed aim was to secure local support within nine previously assigned areas. However, representatives from two of these areas had already communicated such clear negative signals that support was highly unlikely to be forthcoming. In the seven other areas, the plan was to set up a structure of two consecutive meetings with the farmers to dispel uncertainties (first) and to come to a collective decision (second). This is an approach similar to that of the process initiated by the contracted BoerenNatuur in Friesland, but with an obvious difference in participating actors and their responsibilities. Brabant clearly cannot profit from a network of actors specifically focussed on managing wintering geese that has developed over the last decades in Friesland. The meetings in Brabant were organized by the provincial authorities and chaired by a representative of the agricultural organization. What was striking about the meetings we attended was the scepticism and aggressiveness of the audience towards the representative of the provincial government, which clearly reflected a history of built-up distrust that preceded the issue at stake. While most of the farmers we interviewed subscribed to the overall aim of the Fauna Management Framework, the inability of the provincial government to answer concrete questions, which were related to the indemnification of costs and damages as well as their duration, did nothing to dispel the suspicions on the side of the farmers that the state would renege on its promises as it had done many times before. The Brabant authorities were clearly handicapped by the incompleteness of the PFFM for which they tried to find support, although by this time the Minister of Agriculture had officially announced that participation in the PFFM would not have negative consequences for future practices.

Meetings in Brabant were held in only three of the seven area areas. The main reason for this was lack of time, caused by the deadline of June 2005. Since the deadline of June 2005 was rapidly approaching, the provincial authorities, under strong pressure from the Ministry of Agriculture, which served as process manager, issued the order to speed up the process. As a result, instead of setting up face-to-face meetings with those involved, the farmers in question were simply sent a brochure that covered all the information the Brabant authorities deemed necessary while the process of co-decision making was reduced to the official codetermination process surrounding provincial plan making. As a result, Brabant earmarked only 1,600 hectares in just two of the nine initially selected areas as foraging areas. In only one of these areas was the deliberative procedure followed to create collective support.

2.3.6 State of Affairs in April 2006

At the time of writing (April 2006), the Minister of Agriculture has reported back to the House of Representatives of the Dutch Parliament with the message that 100,000 hectares have been made available for wintering geese. It appears that the implementation of the PFFM has been successful and has managed to generate enough support in the twelve provinces. However, some critical remarks need to be made on the situation. The 100,000 hectares are, in part, the result of the initiative of the nature conservation agencies that delineated 40,000 hectares in their care as nature reserves. This is an extra effort that will not add to the costs of the PFFM as these agencies are financed through other means. Although the 60,000 hectares demarcated by the provinces can be regarded as a success in light of the initial resistance to the PFFM, we must, however, not forget that the procedural remedy to the initially top–down approach was focused on acquiring support in previously delineated areas *ex post*. Under the assumption that farmers within the foraging areas stand to gain, which is evidently the perception in Friesland, an attempt was made to convince the resistant agricultural communities in the other provinces that they stood to gain by participation in the PFFM. The realization that those farmers delineated outside the foraging areas have not, in fact, been involved in the process makes it difficult to legitimate the overall PFFM. It is not unthinkable that those initially left out of the process will start to realize that they do indeed stand to gain if located in a foraging area, especially when their efforts to dispel the wintering populations proves to be unfruitful and the farmers suffer damages that are not compensated.

Although 100,000 hectares have been delineated and the PFFM has taken effect, in all likelihood this will not end the discussions regarding wintering geese. The SAN packages have still not been approved, and until they are, the PFFM is not financed by EU money. Consequently, the €14 million for compensatory damages will have to be paid by the Dutch taxpayer. Most of the provinces have assigned their quotas and have, as a consequence, distributed their financial means. Only provinces like Brabant, which did not succeed in appointing its total quota, have room to manoeuvre in delineating their total quota allotted when the larger agricultural community becomes convinced that participating means benefiting, rather than losing when left out. The other provinces, with the exception of Friesland, might be confronted with an agricultural community that demands participation in the PFFM, but will have no financial room for extra delineation.

The uncertainties permeating the PFFM remain. Although some of the uncertainties have diminished, such as the effects on future agricultural practices, other uncertainties prevail or have surfaced, e.g. the effectiveness of the 'stick and carrot' strategy of the 80,000 hectares. During the implementation of the PFFM, contra expertise from Birdlife indicated the necessity of 150,000 hectares to safeguard the wintering populations, which will have considerable consequences when this proves to be accurate. An elaborate evaluation process has been set up for which Alterra, a Dutch scientific research institute focused on green space, has been contracted. Alterra will analyse and evaluate both the ecological and the social aspects of the policy.

2.4 Reflection and Discussion

The purpose of this chapter is not so much to assess the legitimacy of the PFFM in the form of a normative judgment as to whether it is legitimate or not. In this chapter we try to illustrate some of the problems and possibilities of legitimacy production in the hope of contributing to how to deal with the process of making necessary tradeoffs between the input, throughput and output requirements of legitimacy. To accomplish this, we need to evaluate both the efficacy of the policy (output) as well as the processes that led up to the decision and implementation (input and throughput).

As indicated earlier in this chapter, acceptance can be regarded as a measure of legitimacy. However, the absence of acceptance and the resistance to decisions or policy does not automatically imply an absence of legitimacy. The initial resistance to the PFFM should, therefore, not automatically be interpreted as deficient in legitimacy. The PFFM provides for a redistribution of resources from some agents to others. Unpopular policy or decisions are not improbable in these situations, even if these envisage providing an often necessary and urgent remedy to societal problems. The intended result of the PFFM was and still is; (i) providing a durable solution to the agricultural community while at the same time, (ii) safeguarding the geese populations, and (iii) keeping the costs of indemnification in check. This intended 'output' is predominantly based on substantive sources: scientific arguments provided by the Ministry of Agriculture in its (un)surprisingly traditional top–down role! However, the limits of rationality, exemplified by the many scientific and practical uncertainties embedded in the proposed solutions, combined with a multitude of actors who have different core interests, created a situation in which resistance to the PFFM was likely to occur. The lack of legitimacy is, in this case, not due to the actual resistance to the PFFM, but due to a lack of reflexivity on the part of the parties responsible for the PFFM. This is surprising as the recent shift of strategy, as set out in the policy report 'Nature for People, People for Nature', suggests a more bottom–up approach that is 'strongly anchored in the hearts of the people and in the decisions of citizens, entrepreneurs, social organizations, and local authorities' (LNV 2000:18).

Although the Ministry of Agriculture should have and could have been more in tune with the sentiments in the field, the apparent lack of reflexivity could also be ascribed to the 'multilevel' conditions of this particular case. If the case of wintering geese is anything to go by, we see that Dutch nature policy-making is a multilevel governance setting, albeit to a lesser extent and in different guises than is generally thought, for even in this relatively new policy domain the impact of supranational policy making appears limited. While conventionally described as a process whereby national regulatory agents are increasingly restricted in scope and depth by supra-national polities and agents like the EU, it is obvious, as is demonstrated in this chapter, that national authorities remain in control of content and strategy. At the same time, the importance of other public and private agents in the policy process are apparent as well. Although the Ministry of Agriculture has taken the initiative in defining the solution to the problem, the provincial authorities have been legally responsible for the implementation of nature conservation policy and subsequently for dealing with wildlife damage since 2002. This new strategy, in which the Ministry of Agriculture proposes action but leaves the implementation to lower administrative bodies, can explain, to a certain extent, the lack of reflexivity. The redistribution of responsibilities seems to have created a transient situation where the different actors are probing to find out who is responsible for what. The situation in Brabant is illuminating in this respect, as they resisted implementing the first phase of the PFFM but ultimately decided to tag along for the ride and delineated foraging areas in a top–down process. The Friesian authorities are the exception to the rule in that they never initiated a top–down attempt, aware of the resistance this approach would cause.

In terms of shifts between the different modes of the production of procedural legitimacy – input, output end throughput legitimacy – this case is enlightening. What started out as a policy that largely leant on output legitimacy was over time confronted with a need to install more demanding forms of procedural legitimacy, namely input and throughput legitimacy. As was demonstrated by the unwillingness of farmers to accept the expert judgments of environmentalists and policy makers, substantive sources of legitimacy, in particular scientific expertise, have gradually lost their self-evident nature. This is both the result of reasons that are internal to the environmental sciences (paradigmatic pluralism) and of the increasing politicization of environmental planning. The material presented here also provides some backing for Scharpf's (2004) claim that not each and every decision raises similar legitimacy requirements. Dutch policies concerning wintering geese remained largely uncontroversial as long as each and every farmer could claim indemnification. The situation started to change when the state changed its policy to a distribution regime, which, in fact, introduced scarcity. After all, participating in the PFFM is supposed to provide a farmer with a durable solution, which entails clarity about the amount of indemnification and clear expectations of a farmer's responsibility. Although able to dispel visiting geese, the farmer outside the established foraging area received no indemnification after suffering damages.

This situation of scarcity is best illustrated by the situation in Friesland, where support was found for 50,000 hectares whereas the provincial authorities only had 30,000 hectares to distribute. This confronted the provincial authorities with the task of making choices as to which farmers stand to gain, by participating in the PFFM, and which farmers are excluded. It seems, however, that this is only the case within Friesland. Outside Friesland, the uncertainties and the top–down approach contained in the PFFM had the opposite effect. The advantages of participating were not seen, certainly not in the first phase of implementation. This resistance is, however, not necessarily a vote for the option of dispelling without indemnification. Rather, it could be understood as a dismissal of the aggregate of the PFFM. Thus, no matter where you are located, inside or outside the foraging site, the perception is that you stand to lose.

The procedural remedy that was undertaken to counteract the resistance to the top–down policy proved to be both successful as well as unsuccessful. The characteristics that allowed for a productive process in Friesland were not present in

Brabant, but, more importantly, were also not allowed to develop. The procedural attempt to gain support and secure legitimacy *ex post* proved to be only partially successful. The time constraint of the deadline, the uncertainties of the PFFM and the lack of a network of actors to fall back on, proved to hinder the process of acquiring support. The provincial authorities of Brabant were caught between a rock and a hard place. Although the provincial authorities were aware of their responsibility in implementing the PFFM, it did not want to assign the areas without the support of those communities that are ultimately responsible for the success. The need for a distributive solution was not perceived at the provincial level, but the national strategy required active involvement. At the same time, the provincial authorities had been given little or no means to communicate about the PFFM to gain support. The announced strategy of gaining support, articulated at the national level, proves to be understood by the agricultural communities in Brabant as lip service. The Ministry of Agriculture can be held accountable for assigning a time limited task with many uncertainties that can expect to generate a great deal of resistance. At the same time, the provincial authorities can be held responsible for so easily agreeing to the task assigned to them, especially when one realizes that they have significant responsibilities under the FF-Act.

More work needs to be done to assess what the relationship is between the shifts in governance and the shifts in legitimacy production that can be observed in this case. The shifts we can observe in this particular case are mostly reactions to failed output attempts rather than conscious choices to produce legitimacy from start to end. What seems crucial is the ascendance of a certain measure of reflexivity on the side of public authorities and policy makers. For only if the limits of rationality are fully recognized – because of planning failures or because of bottom up resistance ('the weapons of the weak') – and hence a search for knowledgeable agents is initiated with whom the burden of governance could be shared, is there a shift from substantial to procedural legitimacy to begin with. Only if procedural modes of legitimacy production have become predominant, will it be possible to observe shifts between input, output and/or throughput legitimacy.

However, this does not (yet) tell us anything about the reasons behind the shifts between the three dimensions of legitimacy. It could well be the case that upon closer consideration of the case material presented here, intermediate variables, such as available time (the deadline) or the degree of centralization and/or decentralization, appear to be more important. More details are also needed in order to assess these shifts normatively. Any assessment will have to be multidimensional since we live in normatively complex world in which none of the available normative criteria can cover all of our legitimate moral intuitions (Bader and Engelen 2003). To greatly simplify this, we will need both prudential and moral criteria and will have to be perceptive enough to deal with the many trade-offs between them. One such trade-off can be observed in the case material presented above. Because of time constraints, the provincial authorities in Brabant replaced the deliberative consultation process by a much more minimalist and formal consultation procedure. As such, this was a shift from throughput-modes of legitimacy to a more output-based mode of legitimacy. What this implies is that throughput procedures are good in inclusiveness and equal democratic participation but bad in efficiency and effectiveness, while for output-based modes of legitimacy it is the other way around. In future research and analyses, we not only may wish to identify more of these trade-offs, but, as ethicists and political philosophers, ought to search, in a more prospective fashion, for modes of co-decision-making that might soften the starkness of these very same trade-offs.

References

Akkerman, T., M. Hajer and J. Grin (2004). 'The Interactive State: Democratisation from Above?' *Political Studies* 52(1): 82–95.

Bader, V. M. and E.R. Engelen (2003). 'Taking Pluralism Seriously. Arguing for an Institutionalist Turn in Political Philosophy'. *Philosophy & Social Criticism* 29(4): 375–406.

Ebbinge, B.S. (2003). 'Advies aan Faunafonds inzake heropening jacht op Kolgans, Grauwe Gans en Smient'. *Alterra-rapport* 802. Wageningen: Alterra.

Ebbinge, B.S., M. Lok, R. Schrijver, R. Kwak, B. Schuurman and G. Müskens (2003). 'Ganzenopvangbeleid : internationale natuurbescherming in de landbouwpraktijk: van verjagen naar ganzenopvang.' *Alterra-rapport* 792. Wageningen: Alterra.

Ebbinge, B.S. (2004). Ganzenopvang in de Landouwpraktijk. Agri-Monitor: Actuele informatie voor land en tuinbouw. http://www.lei.wur.nl/nl/content/agri-monitor/pdf/feb2004ganzenopvanglandbouwpraktijk.pdf

Faunafonds (2004). Jaarverslag 2004, Periode 1 januari 2004–31 december 2004. Dordrecht

Friedrich, C.J. (1972). *Tradition and Authority*. London: MacMillan.

Joustra, W. (2001). 'Paradijs Gaasterland Telt natuur in punten en niet in hectares.' *Noorderbreedte*. http://www.noorderbreedte.nl/?artikel=603

Groot Bruinderink, G. (1987). 'Begrazing van cultuurgrasland door wilde ganzen.' *De Levende Natuur* 88: 200–204.

Hajer, M. and H. Wagenaar, eds. (2003). *Deliberative Policy Analysis. Understanding Governance in the Network Society*. Cambridge: Cambridge University Press.

Koffijberg, K., B. Voslamber and E. van Winden (1997). 'Ganzen en zwanen in Nederland. Overzicht van pleisterplaatsen in de periode 1985–94.' SOVON Vogelonderzoek Nederland. Beek-Ubbergen

Kuindersma W. and G. Kolkman (2005). Vertrouwen en samenwerking in het experiment Gaasterland. Een procesevaluatie over tien jaar natuurontwikkeling(en). Alterra rapport 1229. Wageningen: Alterra.

LNV (1990). *Ruimte voor ganzen : nota over het ganzenbeleid*. Ministerie van Landbouw, Natuurbeheer en Visserij. Den Haag.

LNV (2000). *Natuur voor mensen, mensen voor natuur: nota natuur, bos en landschap in de 21e eeuw*. Ministerie van Landbouw, Natuurbeheer en Visserij. Den Haag.

Molenaar, J.G. de, D.A. Jonkers, P. Vereijken and G. Kolkman, 2005. EHS-Experiment Gaasterland;2. Effectiviteit Agrarisch Weidevogelbeheer. *Alterra-rapport* 1131.Wageningen, Alterra.

Molenaar, J.G. de, D.A. Jonkers and G. Kolkman, 2006. Meten en wegen in het experiment Gaasterland. Een ecologische evaluatie van agrarisch en particulier natuurbeheer. Alterra-Rapport 1230. Wageningen, Alterra.

Roomen, M.W.J. van, E.A.J. van Winden, K. Koffijberg, B. Voslamber, R. Kleefstra, G. Ottens and SOVON Ganzen- en zwanenwerkgroep (2004). Watervogels in Nederland in 2001/2002. SOVON-monitoringrapport 2004/01, RIZA-rapport BM04.01. SOVON Vogelonderzoek Nederland, Beek-Ubbergen.

Sanders, M.E., W. Geertsema, M.E.A. Broekmeijer, R.I. van Dam, J.G.M. van der Greft-van Rossum and H. Blitterswijk (2004). Evaluation of policies relating to the National Ecological Network and geese protection. Background document for the 2004 Nature Balance. Wageningen, Nature Policy Assessment Office, Wageningen, Planbureaurapporten 11.

Scharpf, F. (1970). *Demokratietheorie zwischen Utopie und Anpassung.* Konstanz: Universitätsverlag.

Scharpf, F. (1999). *Governing in Europe. Effective and Democratic?* Oxford: Oxford University Press.

Scharpf, F. (2004). 'Legitimationskonzepte jenseits des Nationalstaats.' *MPIfG Working Paper* 35. Cologne.

Selnes, T., W. Kuindersma and M. Pleijte (2006). Natuur en gebieden: tussen regels en gebruik; De omgang met bestuurlijke spanningen in drie gebieden. Den Haag, LEI, Rapport 7.06.05.

Strange, S. (1996). *The Retreat of the State: The Diffusion of Power in the World Economy.* Cambridge: Cambridge University Press.

Weber, M. (1972). *Wirtschaft und Gesellschaft; Grundriss der Verstehenden Soziologie.* Tubingen: J.C.B. Mohr (originally published in 1921).

Zouwen, M. van der and J.P.M. van Tatenhove (2002). Implementatie van Europees natuurbeleid in Nederland. Planbureaustudies nr. 1, Natuurplanbureau. Wageningen.

Chapter 3
Vindicating Arrogant Ecologists

Wim Dubbink

3.1 Introduction

In the chapter on *Wintering Geese in the Netherlands*, four stories are skilfully interwoven. The first, a case study on wintering geese, unfolds into three separate stories, each taking place in one of three Dutch provinces: Friesland, Brabant and Limburg. The second is about the development of Dutch nature policy from its inception to the opening years of the new millennium. Then, the author gives an analytical account explaining a new or alternative conceptual framework, which can be used to deal with issues of legitimacy. Finally, this chapter cannot be disconnected from the larger story told in chapter one on historical developments and socio-political circumstances that make the new framework relevant in two ways: as a tool for analyzing cases and as a normative model to be used in policy contexts. Because of the chapter's breadth and depth, I will not even attempt to deal with all the issues raised and the claims made in the work. Instead, I will focus only on the second story – the development of Dutch nature policy – and make only occasional quick excursions to the other stories, especially to the case of the wintering geese.

Dutch nature policy of the eighties and nineties of the last century is portrayed not only as 'arrogant' but also as a typical example of centralist, top down planning. These observations are not particularly positive, especially not in a chapter in which participatory and interactive forms of policy making are associated with democracy and other things thought to be of value. Considering this, it may seem surprising that there are actually nice, reasonable, politically aware, democratic, but perhaps too modest ecologists working for the Ministry of LNV, who were and still are willing to defend this 'arrogant' policy and who disagree with the proclaimed change made in the year 2000 report *Nature for People, People for Nature* (Ministry of Agriculture, Nature Conservation and Fishery 2000). What do these public officials have to say for themselves?

Wim Dubbink
Department of Philosophy, Tilburg University
E-mail: w.dubbink@uvt.nl

J. Keulartz and G. Leistra (eds.), *Legitimacy in European Nature Conservation Policy:
Case Studies in Multilevel Governance*, 47–53. © Springer 2008

3.2 Public Support

Based on personal discussions with these ecologists in the late nineties, a fair reconstruction of their side of the story might run as follows. At the end of the eighties, nature policy finally became something of a priority in the Netherlands. Ever since the effects of industrialization, population growth, and modernization became apparent, the concern for the preservation of nature became an issue in ever-wider circles in the Netherlands. Many people came to understand that something really needed to be done, particularly in the densely populated Netherlands. Compared to neighbouring countries such as Germany and France, the Netherlands had little (terrestrial) nature left, at least in terms of acreage.

A policy aimed at the conservation of nature met with great public support and was backed by all political denominations. This inevitably led to the *Nature Policy Plan* (Ministry of Agriculture, Nature Conservation and Fishery 1990). We would certainly be doing a disservice to history by suggesting that this plan was the result of a *coup d'etat* by a bunch of perverse ecologists. In fact, the Ministry came up with the plan as a response to the concern from both the public and the government. That the Ministry put professional ecologists in strategic positions in the policy process, thus ensuring that the plan would be strongly influenced by ecologists is not surprising and completely defensible. An efficient nature preserving policy must, by definition, be strongly influenced by ecologists. The Delta Works were also put in the hands of hydraulic experts, not under the authority of a discussion group of well meaning amateurs. When the Nature Policy Plan was ready, it was brought before the government and parliament for approval. Every one loved the plan and approved it. In fact, hardly anything was changed or amended in the plan. That cannot be said of many policy proposals, and it is fair to interpret this approval as a sign of the support for the plan.

3.3 An Ambitious Plan

The Nature Policy Plan of 1990 was very ambitious. It was not only about the preservation of what was left of nature but also provided for the expansion of the acreage and for the recovery of lost eco-systems. It also had a very long time span: the complete implementation could take as long as 25 years. The plan is without exaggeration the most drastic infrastructural project in the Netherlands since the Delta Works. It surely outshines other relatively large projects such as the new railroad between Rotterdam and the German hinterland (The '*Betuwelijn*') and, for that reason, deserved the worldwide attention it received. In the 1990's, the Ministry started implementing the plan. This process is now about half completed. Considering its size, the project is still going reasonably well. The *Natuurbalans* (Milieu- en Natuurplanbureau 2005), a yearly auditing report on the status of both nature and nature policy, claims that the acquisition of acreage reserved for nature is still on schedule. Public support for a good nature policy is also still considerable; although the government made serious cuts in all public policy budgets, they have spared the budget reserved for nature policy until now.

Of course, that does not mean that there are or were no (serious) problems or disappointments in the process of implementation. The greatest disappointment is, without doubt, that the environmental policy cannot keep pace with the nature policy. As a result, eco-systems are often difficult to restore because the environmental conditions are simply not good enough. There are also sometimes problems with the locals in particular projects. The attempts to restore peat, for example, have sometimes led to an enormous increase in mosquitoes that harass the surrounding farms and villages, thus straining local support. In other places, farmers have complained about specific measures that they feel threaten their interests. The case of the wintering geese is a good example of this kind of problem. However, the geese case is also an excellent warning not to overstate the problem. As the case makes clear, there were no problems in either Limburg or Friesland. Even if that is not much of an achievement in Limburg, it certainly was in Friesland.

Two things are perhaps even more amazing than the success of the implementation process. Firstly, all the results were gained *without* the use of judicial power. The only two instruments used in the process were the power of persuasion and, of course, (a good deal of) money. Typically, nature policy in the Netherlands does not and never has allowed the use of force to obtain policy goals. In the 1970's, the use of more forceful judicial instruments was once suggested in parliament and the minister promised to study the possibilities, but he never returned to the issue (Ministry of Agriculture and Fishery 1974: 39).

Secondly, the fundamentals of the Nature Policy Plan are still in place. However, they were in jeopardy perhaps once. At the end of the millennium, the Ministry worked on an update of the plan (Dutch law requires an update every eights years), which gave direct human enjoyment of nature a much bigger place in nature policy than the ecologically oriented Nature Policy Plan of 1990 (Ministry of Agriculture, Nature Conservation and Fishery 1999). This plan, however, was never made public and was replaced by the 2000 plan *Nature for People, People for Nature* (Ministry of Agriculture, Nature Conservation and Fishery 2000). This plan certainly cannot be called a major change in policy, but we can clearly see the fundamentals of the old plan, adapted here and there.

3.4 Threats

Looking back at the process from its inception until today, I suspect that many ecologists will not be completely dissatisfied. Many things went wrong perhaps, but some went right. Turning to the future, we have a reason to be optimistic. Public support for nature policy is still intact. When asked to define the most serious threats to a good nature policy, I think that the average ecologist would come up with four issues. Number one would almost certainly be the current state of the environment. Nature policy will always be under pressure without stopping the degradation of the environment.

The ecologist will also be worried about the increasing popularity of interactive policy in all its forms. This is not because he believes in *command and control* per se but because she will have a keen eye for the downside of interactive policy initiatives. This has been empirically documented quite well by Philip Selznick (1949) in his classical study on the Tennessee Valley Authority and boils down to the fact that policy makers can let themselves be taken hostage by local interests. This danger is especially great if the concern for local interests is not considered as a normal part of any liberal democratic policy implementation process but is, instead, glorified as the paradigm of democracy and legitimate policy. When this danger becomes real, interactive policy will in fact end up as the antithesis of democracy. This will be the case when a policy that is approved by a democratically chosen government and supported by the public, is frustrated by a few people (see also Selznick 1949; 1992).

The third problem is the ways in which the Ministry of Agriculture, Nature Conservation and Fishery, in the fight for its own survival, is interfering with nature conservation. The problem goes as follows. The Ministry of Agriculture, Nature Conservation and Fishery is actually a small ministry. It is a public secret that the installation of every new government is awaited with fear and trepidation by the Ministry top brass. Will the new government finally divide up the Ministry between other ministries? Of course, the Ministry itself does everything it can to prevent its liquidation. That is why it is always fighting to increase its responsibilities and its competences. And, of course, that is why the Ministry and the provinces are always at odds with one another, especially since the latter took over some of its responsibilities in a large decentralization operation, which was forced upon it in the nineties. There is always a real danger that this power struggle will damage the process of developing and executing democratic and reasonable policies, which is supposed to be the most important concern of the Ministry. Although not explicitly mentioned in the chapter, the case of the wintering geese clearly illustrates the power struggle between the Ministry and the provinces.

The fourth concern is the continuous flirtation with decentralization in Dutch politics. Of course, any issue must be dealt with at the lowest possible level. With regard to some aspects of nature policy, the national level is the lowest possible level. This puts limits on the possibilities of interactive policy making and processes of participation. However, this is not to say that our ecologist is completely against the new ideas of 'nature for people'. The typical ecologist at the department of Agriculture, Nature Conservation and Fishery is not critical about *Nature for People, People for Nature* because many think that there ought to be more concern for the ways in which ordinary people can and want to enjoy nature. The ecologist's concern is much more specific. According to her, however important the 'nature for people' part of that document is, it should not be a concern of the Ministry. This part of nature policy is something that the Ministry should have left in the hands of the provinces, which are much better suited to taking care of it. The Ministry also should have restricted its activities to the parts of the policy that are of national concern: enhancing the acreage of nature; conservation of species and conservation of eco-systems. For this reason we can still justify an important role for ecologists in *national* nature policy.

3.5 Empirical Differences

If we compare this reconstruction of the ecologists' point of view with the story told in *Wintering Geese*, many differences come to the fore. For example, there is a sharp *empirical* difference of opinion as to the success of the Nature Policy Plan's implementation. According to the chapter, the implementation process almost came to a standstill. However, according to the ecologists, things are well under way. This is not the place to decide this issue, but based upon the conclusions of the *Natuurbalans 2005*, the least we can say is that *Wintering Geese* is overstating its point, especially in relation to the size of the project.

Another empirical difference concerns the question of whether we can characterize the Ministry of LNV's style of governing before the 1990's as *command and control*. The chapter says we can; the ecologists say we cannot. I hold that each side is partially correct. To understand this, we must distinguish between two phases of the policy process: the policy formation process and the policy implementation process. I suppose the chapter is right in suggesting that the formation of nature policy was a typical example of top down planning, and I think that the ecologists may have sometimes been convinced only by their own point of view in those days. Though strangely enough, politically speaking, this was caused by an excessive amount of legitimacy for their policy, not by a lack thereof, because nobody disagreed with them. However, as regards the implementation phase, I do not think that *command and control* is a proper way to typify nature policy. For one thing, there were no instruments to sustain such a policy. Secondly, *command and control* goes very much against the corporatist tradition that is very deeply embedded in the Ministry – even if the Ministry sometimes breaks with that tradition, especially in times of crisis.

One important qualification must be made, however. Even though there cannot be any *command and control* in the real sense, due to lack of accompanying instruments, under particular conditions it may happen that financial instruments, that in principle presuppose voluntary consent, can be regarded by the subsidized party as *command and control*. This happens when the party handing out the money is able to become the dominant party in the process and is thus able to force its own conditions upon other parties. One such circumstance comes about in a situation in which compensation must be paid for damages suffered. Since the damage has already been done, the recipient is in a weak bargaining position. Of course, the situation only becomes worse in cases in which one large subsidizing party is handing out money to many small recipients. Seen in this light, we have reason to look upon the case of the wintering geese as one in which the farmers can experience the government's style as *command and control*, after all. However, we should not look upon the case of the geese as typical of the nature policy implementation process. In the 1990's, it was a rather atypical case.

A third empirical difference concerns the question of whether a transformation of policy started somewhere in the 1990's. Here, I think we must make a distinction between a real change of policy and a change of discourse. It is certainly true that the Ministry changed its *discourse* in the nineties, stressing more and more the fact that it had become a Ministry that valued interactive policymaking and the like. However,

I do not think that a real change of policy style was behind this change. On the one hand – as already mentioned – the Ministry had always had a strong tradition of corporatism. On the other hand, the new discourse did not stop the Ministry from acting in a *command and control* style if it deemed that fit. The subsidy programme (Programma Beheer) which was issued in the late nineties, for example, was rather top down, even though at that time, the Ministry was already constantly talking about an interactive policy (Ministry of Agriculture, Nature Conservation and Fishery 1997). Still, perhaps we should say that a real transformation came about, in one respect, in the nineties. At that time, the Ministry embraced the idea that private parties – in particular farmers and other private land owners – ought to play a more significant role in (implementing) nature policy. This implied a change from a situation in which the state itself (and one or two large NGO's) bought land and conserved it to a situation in which many small landowners received subsidies for certain achievements in the field of nature. The process, which is still going on, was strongly influenced by ideas on privatization and market freedom. Paradoxically, it is exactly this process that will probably lead to more experiences of *command and control* by the newly involved parties. After all, this process will create several situations in which the Ministry can strongly control nature conservation because of its hold over the subsidies, which the recipients not only want to have, but have made investments to get. Consequently, more cases like 'wintering geese' are expected in the future, not less. And if these cases do not come about, we should perhaps not be too relieved, because in such situations we may fear that the government is actually negotiating over the quality of nature with private parties. That is exactly what the *Natuurbalans 2005* mentions as one of the greatest threats to current Dutch nature policy.

3.6 Moral and Political Differences

Between the account in the chapter and the reply by the ecologist, there are also various moral and political theoretical differences. It would be impossible to deal with them all in this short space. Generally speaking, the ecologist does not see a principle problem between democracy and the *Nature Policy Plan*, i.e. *command and control* politics (as long as certain rights are observed) but she does see a problem between democracy and making the legitimacy of the implementation process dependent on acceptance by the parties affected. The geese chapter argues exactly the opposite. It seems to imply that there is something undemocratic and therefore illegitimate about the *command and control* style of governing and it makes acceptance by affected parties a measure of legitimacy. I will make two comments on this controversy. Both of them articulate my basic sympathy with the ecologist's point of view.

Firstly, it seems to me that choosing Weberian theory as a basis for the theory of legitimacy is not neutral to the issue at hand. Trapped by his time, Weber did not believe that practical rationality (i.e. a rational debate on moral and political issues) existed. As such, he had to conclude that as soon as traditional substantial sources of legitimacy dry up, all we can do is to turn to procedural sources of democracy.

Many (modern) thinkers would dispute this inference, including thinkers such as Rawls (1993), Habermas (1981) and Dewey (1925). Interestingly, each of them is optimistic about the possibility of practical rationality in the modern age and each considers democracy itself to be a modern substantive source of legitimacy. I believe that this source can, indeed, be used to defend the Nature Policy Plan as a legitimate plan. All we actually need is the reasonable assumption that future generations have a rightful claim to a natural environment that is at least as beautiful and full of resources as ours.

Secondly, it seems to me that too much emphasis has been put on the acceptability of a policy by those affected as a measure of legitimacy. Surely, when a policy is implemented in a liberal democracy, the rights and concerns of those affected must be taken into account. That is one of the main differences between a liberal democracy and a totalitarian democracy. In the geese chapter, however, it seems that this is the *main* indicator of legitimacy. Indeed, in the study, only the farmers and the various layers of government are presented as actors. In such a scheme, the government is, of course, easily perceived as an external force that must gain some legitimacy in the world. The situation is, however, more complex. The various layers of government have already substantially and procedurally established their legitimacy vis-à-vis the public, i.e. the citizens. As such, the government does not start from scratch. It can already be looked upon as a legitimate actor, which actually stands in danger of *losing* its legitimacy, the more it regards acceptance of policy by a small fragment of the citizenry as the sole measure of legitimacy.

References

Dewey, J. (1925/1984). 'The Public and its Problems.' In J. Dewey and J.A. Boydston (ed.) *Essays, Reviews, Miscellany and 'The Public and its Problems'. The Later Works, Volume 2: 1925–1927g*. Carbondale: Southern Illinois University Press, pp. 235–372.

Habermas, J. (1981). *Theorie des kommunikativen Handelns*. Frankfurt a/M: Suhrkamp.

Ministry of Agriculture and Fishery (1974). *Nota betreffende de relatie tussen landbouw en natuur- en landschapsbehoud. Gemeenschappelijke uitgangspunten voor het beleid inzake de uit een oogpunt van natuur- en landschapsbehoud waardevolle agrarische landschappen*. Tweede Kamer der Staten Generaal, zitting 1974–1975, 13.285, nr. 1 en 2, Den Haag.

Milieu- en Natuurplanbureau (2005). *Natuurbalans 2005*. Bilthoven: SDU.

Ministry of Agriculture, Nature Conservation and Fishery (1990). *National Nature Policy Plan. Regeringsbeslissing*. Tweede Kamer der Staten Generaal, zitting 1989–1990, 21.149. nr. 2–3, The Hague.

Ministry of Agriculture, Nature Conservation and Fishery (1997). *Programma Beheer*. The Hague: LNV.

Ministry of Agriculture, Nature Conservation and Fishery (1999). *Conceptnota Natuur bos en landschap in de 21ste eeuw*. The Hague: Interne notitie LNV.

Ministry of Agriculture, Nature Conservation and Fishery (2000). *Natuur voor mensen, mensen voor natuur. Nota natuur bos en landschap in de 21ste eeuw*. The Hague: LNV.

Rawls, J. (1993/1996). *Political Liberalism*. New York: Columbia University Press.

Selznick, P. (1949). *TVA and the Grass Roots*. Berkeley: University of California Press.

Selznick, P. (1992). *The Moral Commonwealth. Social Theory and the Promise of Community*. Berkeley: University of California Press.

The Great Cormorant (by André Kuenzelmann)

Chapter 4
Legitimacy of Species Management

The Great Cormorant in the EU

Felix Rauschmayer and Vivien Behrens

4.1 Introduction

The conservation of endangered species was institutionalized in Europe in the last quarter of the twentieth century, taking the Bern convention as a starting point. The first EC legal framework for species conservation, the Birds Directive, was put in place 25 years ago. The Habitats Directive, which took up the apparently successful approach for other species, was adopted in 1992, but its implementation into national law is still ongoing.

In the last few centuries, large vertebrates, especially top predators, have become endangered because they need natural resources demanded by humans as well. This is why protecting these species is, in general, linked to conflicts. In fact, for quite a few species, modern conservation measures have been so successful that conflicts about their conservation have grown fast in intensity. This is especially the case with top predators that feed on natural resources also used by humans.

The re-introduction of large vertebrate species as well as their re-colonisation of ancient habitats can be considered a major success of national and European conservation policy, even if many populations are not yet viable in Western Europe. This success is obvious for species that have reached a favourable conservation status, such as the opportunistic and very mobile Great Cormorant (*Phalacrocorax carbo*) (Carss 2003).[1] In such cases, the very success of conservation challenges the continued insistence on conservation, not only due to the conflicts over resources used by humans but also because these populations endanger other species. In the case of the cormorant, certain fish species are now under threat (Koed et al. 2005).

The background for this chapter is the EU research project FRAP (Framework for Biodiversity Reconciliation Action Plans, see www.frap-project.net), which dealt with conflicts between fisheries and the protection of cormorants, grey seals

Felix Rauschmayer
Helmholtz Centre for Environmental Research UFZ
E-mail: felix.rauschmayer@ufz.de

[1] In the text, we always refer specifically to the Great Cormorant, and not to other cormorant populations, such as the Pygmy Cormorant (*Phalacrocorax pygmaeus*).

and otters.[2] The common feature of the seven case studies being analysed in this project is the competition for fish between fishermen (professional and non-professional) and top predators inside the European legal framework. FRAP highlights the shift from species preservation to species management which aims at safeguarding a minimum viable population while preserving conflicting economic and cultural interests. Whereas two case studies (Denmark and Province of Ferrara, Italy) deal specifically with the cormorant, conflicts with this species show up in all other cases (Finland, Sweden, Portugal, but especially Germany and the Czech Republic). Stakeholders in the different regions frequently call for a 'European solution' to the cormorant problem, whereas the main European co-ordination project on the cormorant conflict (INTERCAFE,[3] the successor of REDCAFE (Carss 2003)) almost exclusively favours local handling of the specific situations. This dichotomy puzzled us, so we decided to look into the history of and further possibilities for a European cormorant action plan by interviewing government officials, scientists, and representatives of interest groups in different European countries about their experience with and perspectives on the cormorant situation.

This chapter, which only takes up a small part of the FRAP research, is organised as follows. After an overview on the cormorant situation in Europe and the legal background for its preservation and management, several management plans on different levels are sketched, and the multi-level aspect briefly analysed. Their legitimacy is then analysed along four criteria, with a special focus on the European level. The conclusions focus on the difficulty of achieving institutional innovations in multi-level contexts.

4.2 Cormorants in Europe

Through persecution and the destruction of habitats, the numbers of cormorants fell to a very low level all over Europe in the 1960's and the species was threatened with extinction. In 1979, the EC Birds Directive protecting wild birds was enacted. This protection status as well as the protection of breeding sites and the available food supply have led to rapidly increasing cormorant numbers since the 1980's (Carss 2003).

From about 800 breeding pairs in the Netherlands in the early 1960's, the number of cormorants increased to 150,000 breeding pairs in 1995 (Carss 2003). In 1998, the European anglers association stated that there were slightly more than 700,000 cormorants in Europe (European Alliance of Anglers 1998). Estimates of

---

[2] Additional input comes from the EU Co-ordination Action PATH (Participatory Approaches in Science and Technology).

[3] For further information see the INTERCAFE fact sheet following the link: http://www. intercafeproject.net/factsheet.html (date of access: May 15, 2007)

the population today vary between 500,000 and 3 million birds. The increased numbers led to new (or rather renewed) migration schemes of this opportunistic fishing bird; numbers not only increased in the 1970's in the cormorant nesting countries of Denmark and Netherlands, but also in most other European countries where the cormorants forage during winter or during their migration southward, and where they increasingly build nesting colonies as well. Nesting places seem to be the resource limiting their population. Most adult birds do not have a nesting place and occupy the places of dead cormorants. As a result, limited killing of adult birds will not lower the reproduction rate.

A cormorant eats approximately 0.5 kg of fish per day and does so mostly in water bodies rich in fish, i.e. fishing nets, aquaculture, or lakes and rivers used and stocked for sport fishing. The cormorant's eating habits lead to economic losses, but also to resentment towards species preservation, which protects a culturally (but not biologically) new competitor on fish produced by humans. In their respective countries, but also on the European level, professional and sport fishermen increasingly ask for solutions to this resource conflict.

Any cormorant management plan would have to include a set of measures taken on different levels that can help to secure the fisheries' interests while preserving at least the minimum viable population of the target species. Measures (e.g. hindering the cormorant from fishing at specific places through technical measures or by scaring them away, managing locally specific populations by impeding new colonies, or managing the general population through shooting, egg oiling, nest destruction, etc.) vary in their aim and in cost. From a modelling perspective, the common assumption is that a centrally organised management scheme would be the most efficient and cost-effective.

The current cormorant management measures in Europe consist of many uncoordinated management plans that have to respect the Birds Directive's condition that management actions should not endanger the cormorant population. This situation begs the question whether a pan-European management would improve the situation or not. Every country has its own regulations harmonized to the conditions of the conflict and containing site-specific regulations. In the interviews conducted, it was pointed out that due to the uniqueness of the conflict situations, there is no need for a pan-European management. On the other hand, the cormorant population is pan-European (including the whole Mediterranean basin), which means that an action in one country has implications for other countries and the sum of many actions may have implications for the general viability of the species.

The success of a pan-European management plan will depend on how this plan is designed. If it is merely a set of techniques that are applied randomly, the benefits will not be that high. Coordination between the range states concerning the number of cormorants that are killed and the exchange of knowledge on the measures used can be more effective for assessing the problem and can provide an overview about the development of population numbers. This can help to prevent negative impact on the overall cormorant population through uncontrolled and uncoordinated killing.

4.3 Legal and Institutional Background of Nature and Bird Conservation[4]

The major European-wide and EC laws which stipulate the protection of species are the following:

- Council Directive 79/409/EEC of 2 April 1979 on the Conservation of Wild Birds (the Birds Directive)
- Convention on the Conservation of European Wildlife and Natural Habitats
- Convention on Migratory Species (CMS)
- Council Directive 92/43/EEC of 21 May 1992 on the conservation of natural habitats and of wild fauna and flora (the Habitats Directive)

The Convention and the Directives are substantially linked and overlapping, although they can be seen as formally separate. In 1979, the Birds Directive was adopted, and the Bern as well as the Bonn Convention was made. Three years later, in 1982, the Council Decision on the Bern Convention was made. The Habitats Directive was adopted ten years later. At present, the major relevance of the Bern and the Bonn Conventions (CMS) is the membership of countries beyond EU member states. The main aims of the CMS are to preserve migratory species and to establish cross-boundary cooperation with regard to their conservation, rather than to deal with species management issues. In 2004, 86 countries and the EU signed the Bonn convention. The CMS recommended designing an international cormorant management plan, which took place in 1997, and was prepared by Denmark and the Netherlands. Under the patronage of CMS, further international agreements were created, such as the Agreement on African-Eurasian Waterbirds (AEWA), which thematically would be the right frame for a pan-European cormorant action plan.

4.3.1 Birds and Habitats Directives

The Birds Directive covers all bird species naturally occurring in the European territory of member states. It was adopted in 1979, the same year as the Bern and the Bonn Conventions. It was the first significant legal measure of the European Union in the field of nature conservation. The experience gathered from its implementation strongly affected the content of the Habitats Directive more than ten years later. The point of departure for the Birds Directive is that all birds are protected if they are not explicitly excluded from the protection. Certain activities are prohibited, such as deliberate killing or capture, deliberate destruction of nests and eggs and deliberate disturbance of birds, in particular, during the period of breeding and rearing. With regard to threatened species (Annex I), special protection measures are required. The most important of these measures is the designation of special protection areas,

---

[4] This section is mainly based on Similä et al. (2004).

which are now dealt with in article 6 (paragraphs 2–4) of the Habitats Directive (see below). In Annexes II and III of the Directive, those bird species are listed which may be hunted or traded, despite the abovementioned protection. However, both hunting and trading are still restricted and controlled in specific ways. At present, the Great Cormorant is not included in either list, although it was previously considered a threatened species (Annex I). In fact, the Great Cormorant is the first species that has been taken off the list of threatened species. According to Krämer (2000), the deletion is mainly due to pressure from fishermen. Despite this, the Great Cormorant is protected against hunting and selling under both the Birds Directive and the Bern Convention.

The ORNIS Committee is responsible for the implementation of the Birds Directive. It consists of representatives from the EU member states, from the European Commission, and, in its enlarged version (called ORNIS+) also from different NGOs. Since the cormorant was moved from Annex I of the Birds Directive in 1997, the ORNIS Committee has been continually checking the situation of the species. At a meeting in June 2003, the ORNIS Committee stated that there is no agreement on the need for international management of the cormorant (cp. European Union 2004). This was reiterated in March 2004. The draft of the CMS Action Plan on the Management of Cormorants presented by Denmark and the Netherlands has not been formally adopted. One member of the ORNIS Committee stated in an interview conducted as part of our research that there would be no action plan beyond the draft from the Copenhagen meeting. Consequently, no action plan has been elaborated on by the ORNIS Committee.

The Habitats Directive required that the member states are obliged to designate Special Areas of Conservation (SAC) for the habitat types listed in Annex I and species listed in Annex II. Natura 2000, a European ecological network of special conservation areas, includes, in addition to these sites, all special protection areas classified by the member states pursuant to the Birds Directive. The existence of SAC's limits the application of specific management measures with regard to the cormorant. These limitations are not a main obstacle to cormorant management, though.

4.3.2 Mitigation Measures and Derogations from Nature Conservation Law

The protection of species may cause undesirable effects, such as the effects caused by cormorants to fisheries. In many cases, the extent of these effects can be mitigated by taking certain actions. From a legal point of view, the problem with taking mitigating measures is that a number of them may violate the law. For example, hunting of protected species, destruction of nests or eggs of protected species, and deliberate disturbance of protected species, as well as some hunting methods, are prohibited under nature conservation laws. However, nature conservation laws usually contain derogation rules. Under derogation rules, it is possible to allow measures that would otherwise be in violation of the law. In fact, all major pieces of European and international law contain such rules. Derogation rules for species protection and habitat protection are two different things. Species protection rules in the Bern Convention, the Habitats Directive and the Birds Directive have many similarities. The starting point of these laws is that:

- derogation is allowed only if so decided by a party or member state,
- derogation can be granted only if certain general conditions are met,
- derogation can be granted only for legitimate purposes.

There are two types of general conditions related to derogations from species protection provided in the pieces of nature conservation law studied; one relating to alternative solutions and another relating to effects on the population of the species concerned. With regard to the first type of conditions, the Bern Convention provides that derogation may be allowed only if there is not an alternative solution. The equivalent expression used in the Habitats Directive is 'satisfactory alternative' and in the Birds Directive 'other satisfactory solution'. With regard to the second condition, the Bern Convention stipulates that derogation must not be detrimental to the survival of the population concerned. The equivalent expression used in the Habitats Directive is as follows: 'detrimental to the maintenance of the populations of the species concerned at a favourable conservation status in their natural range'. This type of condition does not exist in the Birds Directive.

The 'legitimate purpose' in the Habitats Directive, which has been translated into national laws, relevant to cormorants, is the prevention of serious damage, in particular to crops, livestock, forests, fisheries and water, and other types of property. It is not possible to make an in-depth analysis here of what 'serious damage' to fisheries as a type of property really means. However, it can be assumed that seriousness is measured in economic terms.

4.4 Cormorant Management Plans

Every country deals with its own legislation or regulation on the cormorant issue. While the cormorant is protected in all EU member states due to the Birds Directive, some national or regional management plans provide for rather large-scale killing to mitigate the conflicts between fisheries and cormorants, such as in France. Shooting allowances may be linked to specific sites (e.g. Sweden, Poland, Italy, Denmark), to fixed dates (e.g. Romania, Estonia), or to predefined quotas (e.g. France, UK, Slovenia). In other states, killing or disturbing the bird is illegal, e.g. Belgium and the Netherlands (Carss 2003: 110–113). The general tendency is towards more management of the cormorant with an increase in killing allowances or quotas.

In the interviews conducted, some stakeholders expressed a preference for national solutions with regard to the European level. It was pointed out that problems are local and, consequently, they should be solved on a local level. It was also stated that the Netherlands and Denmark should take action and control the breeding numbers. It became clear that a common action plan without the co-operation of these two key countries would not be effective.

Denmark and the Netherlands play a key role in the issue because they contain the most important breeding colonies. For any population control, these countries would have to be involved, as the availability of nesting sites together with the breeding success seem to be key factors to any population management. Some of the interviewees pointed out that neither of these two countries was interested in international management. In the Netherlands, the cormorant is not seen as a problem due to the high importance put on nature conservation. Most of the Dutch breeding colonies are on private land, so the intervention possibilities regarding population control are limited. In Denmark, the national position is evolving towards more management and less preservation. Other European states like Germany, France, Italy and Austria are more responsive to the fisheries interests, so they are in favour of a more active management appropriate to mitigating the respective conflicts.

Concerned non-governmental organisations are fishermen and anglers associations and environmental organisations. Mainly inland fisheries at the local or regional level perceive a large amount of damage to their fisheries as being due to the cormorant and they claim the need for international management. Such management in their eyes usually implies a huge reduction in size of the population. Nature conservationists, on the other hand, come out against such international management. Other forms of co-operation, such as data or knowledge exchange, are typically welcome by all actors, but are often seen apart from a co-ordinated management plan.

4.4.1 National and Sub-national Management Plans

This chapter concentrates on three exemplified regions or countries with comprehensive management plans to guide conflict mitigation; Denmark, the Italian Province of Ferrara, and France. Species-specific plans define the management goals and the instruments used to achieve these goals. In all those processes leading to management plans, fisheries and nature conservation interests have been considered through participation of NGO representatives in specific advisory bodies. In the following section, the plans are described.

In Denmark, a far more comprehensive action plan for the great cormorant has been adopted than in the two other cases. This is due to the large population of cormorants in Denmark, and the resulting high pressure on the fisheries. The cormorant management plan is a major policy instrument, where the direction of state policy on the conflict is formulated. All three management plans are of particular relevance with respect to 'protective' hunting and other measures to control the size and distribution of the population of protected species. In the province of Ferrara, Italy, the action plan for the cormorant has more limited relevance in guiding the mitigation of the conflicts. In France, the main activities are directed towards culling in order to achieve a target wintering population.

In **Denmark**, the cormorant management plan is the key instrument in managing the conflict. The first management plan was adopted in 1992. In 1994, it was reviewed and the latest version of the plan was adopted in 2002. The 2002 Management Plan was drafted by the Danish Forest and Nature Agency and went through a hearing process where relevant authorities, organizations and interest groups had the opportunity to make comments. After this, the national Game Management Board approved the plan. The Ministry of the Environment has the highest responsibility in the plan. The overall objective of the management plan (2002) is to ensure that the size and distribution of the population does not cause unacceptable inconvenience for fisheries while simultaneously taking into consideration the protection and survival of the cormorants as a species breeding in Denmark. The plan is comprehensive with respect to all basic management measures, and while it has no formal legal effects, the various public authorities follow it. The most used conflict mitigation measures are oiling of eggs to reduce hatchings and scaring cormorants away when they are establishing new colonies by removing nests, making noise or killing a limited number of the birds. In addition, owners of active, fixed commercial fishing gear are allowed to kill cormorants within 1000 m of their gear as preventive damage. Furthermore, experimental hunting is ongoing in a limited and defined area, with a limited number of hunters and a maximum quota on cormorants in order to investigate whether hunting is an effective mitigation measure.

In **Italy**, species are protected under the national-level framework law on the protection of wild fauna and hunting. The regions in Italy have their own laws implementing the national laws in a more detailed manner. Provinces, which are sub-national and sub-regional administrative units, apply the regional regulations. No cormorant species belongs to the list of game species. In theory, however, derogation of the hunting ban could be possible according to the Italian laws on the conditions stipulated in the Birds Directive. The Provinces suggest derogations that the national institute for wildlife then assesses and approves or rejects. The population reduction plans, if such are suggested, are included in the provincial game management plans. There are population reduction plans also for cormorants, but they allow very limited direct killing, e.g. 50 individuals per year in the province of Ferrara. Game management planning follows the Italian administrative structure from the national to the regional level and then to the local (province) level. Provincial hunting areas as well as national parks and other large protection areas have their own game management plans that regulate hunting in that area. These local level plans are checked and coordinated at the provincial level in the province's game management plans, and are then included into the regional level plans. While the cormorant is no game bird, it, nevertheless, has been handled in Ferrara through the *consulta di caccia*, the hunting council, assembling all main stakeholders in the cormorant conflict (except animal rights activists). While the disagreements about numbers and feeding of cormorants have been quite substantial, the council has succeeded in maintaining the conflict at a low level. In the neighbouring province of Ravenna, though, several fishermen claiming compensation for damages due to cormorants took their case to court.

France is the most important area in Europe for shooting cormorants. In 1992, French hunters started to shoot cormorants at a few sites at fish ponds, but also on open waters. These shootings increased from 4,500 shot birds (quota = 6,916 birds) in 1996–97 (Marion 2003) to 25,000 birds in 2001–02. Since 1994, the Ministry of the Environment annually issues a departmental order in accordance with the advisory opinion of the National Council for the Protection of Nature. With this order, prefects on the departmental level are authorized to issue shooting licences to owners or tenants of fisheries. The prefects are assisted by departmental committees in which all parties concerned are represented. The shooting licences are compulsory. The shooting season goes from September to February or March. Despite the efforts that the French government has taken to reduce the cormorant population in the last decade, it has only been successful in slowing down the cormorant's population increase. As in Ravenna, fishermen also went to court and asked for large sums of compensation (more than €1 million). The fishermen's case has reached the *Conseil d'État*, the highest jurisdictional level. This compensation claim, though, stems from the period before the state tried to reduce the wintering population.

4.4.2 The Efforts Towards a European Management Plan

The European dimension of the conflict is obvious as different countries are affected by the same birds (migratory species, with breeding colonies in northern Europe and wintering in southern Europe). As such, it is not astonishing to see that the conflict has been dealt with at the European level. At the 1994 meeting of the Scientific Council of the Convention of Migratory Species (CMS or Bonn Convention) in Nairobi, the cormorant issue was discussed. At the conclusion of the meeting, the Conference of the Parties (CoP) to the CMS recommended the elaboration of a special cormorant management plan. Denmark and the Netherlands were invited to draft a pan-European action plan for the management of the cormorant.

In 1996, an expert meeting took place in Lelystad (The Netherlands), where the current situation was discussed and the latest research outcomes were presented. All available knowledge on the subject was collected, and several management options were considered. These two meetings provided the basis for the final management plan (Van Eerden 1996). In 1997, at the next meeting of the CMS in Geneva, Denmark and the Netherlands were reminded to take the necessary steps and involve the affected states in an international management plan and to finalise it before the end of 1997. The following step was an expert meeting in Copenhagen in 1997, where the drafted plan was discussed by representatives of the concerned states including non-governmental organisations from different levels and interests.

The drafted plan is mainly a set of different measures that can be used by the countries to reduce the damage caused by cormorants. Measures included have also proven useful and feasible in mitigating the conflict. The measures are divided into four main categories:

- technical/physical measures taken outside the breeding colonies
- technical/physical measures within the breeding colonies
- financial compensation measures
- monitoring.

In the first two categories, lethal or non-lethal measures are categorized. It is also noted that measures in breeding colonies have to be in accordance with EU and international legislation. In the preface, the plan clearly points out that the substantial reduction of cormorant numbers will not necessarily lead to a reduction in the amount of conflict in problem areas because killed cormorants will be replaced by other cormorants (Bonn Convention 1997). No clear aim (such as a predefined damage or population reduction) was agreed upon.

The proposed measures are currently applied in several European countries, but this management plan does not seem to mitigate the conflict, especially for the most concerned countries in Europe. In fact, the set of measures proposed provides possibilities that have to be carefully chosen and combined to achieve conflict mitigation. European-wide population management cannot be achieved through a mere listing of measures, and, in fact, was not intended by the main actors.

The representatives of the German Federal Government and the European Inland Fisheries Advisory Commission (EIFAC) did not support the draft management plan. The (then conservative) German government wanted to reduce the cormorant population by 25% annually in order to resolve more effectively the conflicts between fisheries and cormorants, especially in Germany.[5] EIFAC supported this German position and argued that the interests of fisheries were neglected while bird conservation aspects were given priority.

After this discussion, Denmark and the Netherlands finalised their work and sent the plan to the CMS in 1998. However, this plan has never been implemented, and the cormorant issue has never again been taken up within the CoP meetings of the CMS or within its relevant 'daughter' convention, the African-Eurasian Waterbirds Agreement (AEWA).

In 2002, the European Anglers Alliance (EAA) initiated another meeting in Strasburg to draw attention to the cormorant issue once again. Recent research on cormorant populations, damages, and management measures was presented and discussed. It was pointed out by a French speaker that a European harmonisation of regulations is necessary to solve the problem of damage to fisheries. A recommendation to the European Commission was adopted and agreed on by the participants and sent to the EU. However, no action was taken because the ORNIS Committee did not see the need for an international management plan.

Today, almost all European states have their own national or local legislation and regulation on the treatment of cormorants to mitigate the conflicts between these

---

[5] Germany changed its position quite drastically towards more species preservation as soon as the Greens started to participate in federal government in 1998. Other countries, however, have been continually evolving their positions towards more population management and more and more killing of cormorants.

interests (Carss 2003). These regulations contain different measures, which range from protecting the species with very little management (e.g. The Netherlands) to shooting large numbers of birds (e.g. roughly 30,000 in France in 2003/04).

4.4.3 Analysing the Levels of Cormorant Management

The cormorant management issue clearly plays out at various levels. The local level is the place where the damage to fisheries occurs and where mitigation measures are implemented. The material specificity of local sites often asks for locally specific technical mitigation measures. According to the main European Co-ordination Action on the cormorant issue, the INTERCAFE project, this level is the most important in mitigating the conflicts. This position is particularly relevant considering that a general reduction of the cormorant population is impossible. However, even with a heavily reduced population, for example, a population reduced by 50%, local conflicts would still be numerous.

The regional or state level, according to the institutional setting in the respective country, can determine the political style of conflict handling (e.g. the differences between the neighbouring provinces of Ferrara and Ravenna in Italy), or even design specific economic or institutionalised mitigation measures (e.g. the cormorant enactments interpreting the Birds Directive derogation possibilities in German *Länder*, see Thum 2005, 2004).

The national level represents the countries' interests in international bodies such as the ORNIS Committee or the CMS and, in some countries, determines the national policy concerning the cormorant, such as in France. It can be observed that national governments tend to be representative of a balance of positions of 'their' stakeholders. The strength of nature conservation NGOs is in line with the Dutch government position, as is the relative strength of natural resource users such as fishermen and hunters with the French government position.

Finally, the international level is characterised, on the one hand, by the EU and its relevant bodies and, on the other, by the CMS and AEWA. Activities on this level are called for by many stakeholders, but the need for agreement, especially with the Netherlands and Denmark, render action towards a European management plan difficult. The CMS (or its 'daughter' convention AEWA), which initialised the first pan-European Cormorant Action Plan, is apparently not judged relevant for the resolution of the conflict. The focus on the EU, and here on the ORNIS Committee, excludes some countries concerned by the cormorant migration, such as Russia or Morocco, while aiming at a stronger frame for such a plan (Fig. 4.1).

Different stakeholders are active on the respective levels. While the fishery organisations are usually present at all levels, science and species conservation tend to focus their activities on the national and international level. Their local impact is high, though, because of the high weight given to scientific information and because of the existing regulations, which still rather preserve the cormorant. The weight of the individual stakeholders differs according to the specific context. Whereas the

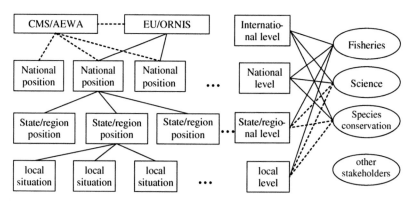

Fig. 4.1 Links leading to European cormorant action plan

fisheries complain about the low attention paid to their interests, for example, in Denmark and Sweden, the national positions in France, and during the conservative government in Germany, reflect much more the fisheries' points of view.

Science has mostly been used to support species conservation interests, which is related to the fact that some scientists understand themselves as 'bird lovers', for example. Authorities have complained that no comprehensive scientific model exists proving the migration and locally specific damages of cormorants. Such a model is thought essential to convince reluctant actors of the necessity of a European-level management.

4.5 Legitimacy of Cormorant Management

In the cormorant case, two major changes are occurring over time that cause the conflict to be handled differently:

- The escalation of the conflict due to strong population growth (itself due to environmental change and successful species preservation) as well as changing human behaviour (increase in sports fishery) and animal behaviour (more migration, different hunting techniques).
- A subsequent shift from species preservation to species management demanded by the fisheries and mostly accepted to some degree by all stakeholders.

Both changes challenge the old institutional situation, which was characterised by mere species preservation. These changes not only challenge its legitimacy, but also impose new demands on information and its management and on social dynamics of the conflict partners. The changes also increase the costs of the conflict and its mitigation.

When judging the overall soundness of a decision to be taken, legitimacy is one main aspect to be considered. Turning one's attention to legitimacy may result in three advantages:

- The legitimacy of decisions may be considered a value in itself (normative character of legitimacy),
- The process leading to legitimate decisions may improve the quality of the decision (substantive character of legitimacy), and
- Decisions that are considered legitimate are usually easy to implement (instrumental character of legitimacy).

Other factors as well have these three characteristics. Improved information management can also lead to improved implementation, substantially better decisions, and have its inherent value (when adopting a rationalist perspective). The same can be said for social dynamics which work well, and for cost-efficient processes and decisions (cp. Wittmer et al. 2006).

The selection of an appropriate decision process is important, as each process has different advantages and disadvantages according to the characteristics of the conflict. While mere disposal of credible information might be enough in low-intensity conflicts about data, highly elaborate and costly processes may be the adequate means to deal with high-intensity conflicts about values and interests. Particular attention should be paid to information management; while scientific knowledge is often the first base of biodiversity action plans, local knowledge frequently does not find its way into the decision processes on action plans. This one-sidedness, apart from substantive shortcomings, can have severe implications on the legitimacy of the plan and hence on its implementation.

4.5.1 Criteria for Evaluating the Legitimacy of Decisions

If the legitimacy of a decision is doubtful, then it is also doubtful that this decision is compatible with generally agreed principles. A decision may be legitimate if it is functional for the well-being of the persons affected or, alternatively, if these persons (or a majority thereof, if procedural fairness is guaranteed) freely express their agreement with the decision. The debate on 'ecological dictatorship' shows that there may be a high tension between these two differing models of legitimacy (Kauffmann 1999). In each society, rules and procedures exist for the resolution of conflicts and for what is considered as fair in terms of process as well as in terms of outcome. A decision is often considered legitimate if it is compatible with the prevailing set of rules in a society. This understanding of legitimacy combines the first two reasons given for legitimacy under the assumption that the prevailing set of rules is functional for the well-being of the persons affected and that these persons would accept any decision that is compatible with this set of rules. If a decision-making process is incompatible with the prevailing set of rules in a society, it might fail for this reason.

In a first approximation, we define a conflict resolution process as legitimate if it is perceived as complying with the formal and informal procedures recognised as adequate in the respective context by all parties affected. In assessing legitimacy, one has to take into consideration that decision-making contexts differ widely:

- Standard decisions are usually taken by administers and are judged legitimate if the administrative procedures follow the legal and paralegal rules.
- Decision contexts involving new elements of knowledge affect new groups of people, or – where uncertain outcomes are related to the decision options – demand for non-standard decision procedures. Such new procedures may require the involvement of knowledge holders such as scientists, stakeholders, or the general public in order to be perceived as legitimate.

Our context, the management of a rapidly-growing species of no hunting interest, demands new knowledge, affects new groups of people, has decision options with uncertain outcomes, and therefore demands the participation of people outside administration. It is an open question, though, which process is appropriate and on which level it should be conducted. The efforts towards a European management plan have been dominated by national authorities with little participation of NGOs and with no participation of the general public.

The acceptance of the process and its outcome by the local and broader community, including the political and administrative actors involved, is contingent on the legal and political as well as on the cultural framework. According to the legal framework, certain aspects might be formally regulated or open to debate. Two especially important aspects in the context of participation are, first, who holds the rule-making authority, including the decision on whether participation is granted voluntarily by an agency or an actor or whether it constitutes a right of certain groups or individuals. Second, the decision-making power or authority includes aspects such as how binding the outcome is and who can be held accountable for it (Steelman and Ascher 1997). A closely related aspect of accountability is the question: Who has a right to participate? Often, participation is abused as a means to veil accountability and not used for substantive or normative reasons (Stirling 2006).

Participation can range from the inhabitants of an area, i.e. those directly affected, to representatives of predefined interest groups (Johnson and Wilson 2000; Davies et al. 2005). This dimension of a good decision process concerns questions such as whether all relevant interests are represented adequately and whether the procedure permits protection or, ideally, enhancement of the interests involved. Inclusion per se is by no means sufficient. The issue is whether all relevant interests and affected stakeholders are known, included and/or represented in a way to assure their equitable participation in the process. At the same time, the power relations between different stakeholders need to be taken into account. It is important to keep in mind that social dynamics do not simply disappear by inviting all relevant interest groups to participate. A decision-making procedure can enforce or counteract prevailing differences in power. This also includes questions such as: Are the least powerful represented or are they even known about? Are the rules such that they can articulate their interests? (cp. Johnson and Wilson 2000).

It is by no means simple to identify all the interests involved in a specific conflict, especially when taking into account that many environmental decisions have far reaching consequences with regard to time and space. 'Global warming' or 'loss of biodiversity' are just the most evident examples. It is still more difficult to represent all interests, even the interests of those who cannot express them. There is no such thing as the 'right' way to represent future generations or other species, but decision procedures allow for such a representation to a different extent. In conflicts where such considerations are important to some of the participants, reflections on representation of interests that cannot be expressed by those who have the interests should be included in the selection of an appropriate method.

Obviously, trade-offs between the quality of a decision-support process and the directness of participation will occur. Where participation is direct, as in referenda, at least all those eligible to vote can be involved in the decision. However, the amount of information that can be managed in a referendum is minimal (cp. Steelman and Ascher 1997). In the face of doubts on representativeness, the transparency of decision-making procedures can contribute to the perceived legitimacy of the decision-making process.

Based on the above, we suggest the following set of criteria for evaluating the legitimacy of a decision:

1. **Legal compatibility**: Are the procedure and the proposed outcome compatible with existing legislation?
2. **Accountability**: Is someone held accountable for the decision and its outcome? Is it clear who?
3. **Interest representation**: Are all the relevant interests included or at least represented?
4. **Transparency**: Are rules and assumptions transparent to insiders and outsiders?

These aspects taken together determine the risk of a decision being rejected as illegitimate. They obviously depend heavily on the political and legal context.

4.5.2 Legitimacy Analysis of Existing Action Plans

These four legitimacy criteria are now used to highlight specific aspects of cormorant management plans before turning specifically to the Danish case and the European level.

1. **Legal compatibility** At least in the EU, the Birds Directive gives the general framework for cormorant management plans. Legal analysis as well as the reactions of interviewees lead us to believe, though, that this framework is not very concrete and that administrative procedures can counter the procedure that derogations have to be agreed upon by the European Commission. The German *Länder*, for example, pass enactments for cormorant regulation, the compatibility of which with national and European law cannot be challenged in court (Thum 2005). Due to defective knowledge on the population structure, it is very hard for the EC to state that a planned management measure endangers the European cormorant population. As the action plans generally are elaborated under the direction of public authorities, legal compatibility should not pose a major problem.

2. **Accountability** For the existing plans, this point is more crucial than legal compatibility. During the elaboration and implementation phase, existing institutions are confronted with the institutional setting preponderant in Europe. In most countries, nature protection agencies are responsible for species preservation, and resource-use agencies (such as fisheries authorities) regulate (and sometimes stimulate) human use of natural resources. None of the agencies is accustomed to species management aimed at safeguarding a minimum viable population while preserving economic and cultural interests conflicting with the resource use of the target population. Both sides have to be accountable for such species management, though, if one does not want to introduce a new type of authority. Consequently, both sides will have to be open to take a new role, and both sides will need social trust if the conflicts are to be mitigated.

3. **Interest representation** The shift from species preservation to species management is accompanied by a shift in interest consideration. For example, in Denmark, nature conservation interests prevailed in the earlier management plans, while the fisheries interests gained more weight in the last phase. In most countries that we considered, nearly all interests were included, whereas the weight given to specific interests mostly depended on political and practical considerations. In most cases, authorities worked with regional or state representatives of fisheries and angling associations rather than with local fishermen. This indirect interest representation apparently made conflict resolution easier, but locals still had the impression that they were not being heard in the decision process.

4. **Transparency** Elaborating management plans was not done with formal algorithms to evaluate different strategies, and all were framed in rather informal meetings. The decisions that stakeholders had to consider were not explicit, but rather decided upon by the process's organising authority. Scientific knowledge was given priority over local knowledge without making this decision and its implications clear to all participants.

Within the field of legitimacy, accountability and inclusion of interests pose the major problem for national and sub-national plans. In practice, though, mistrust in nature protection agencies and a perceived detrimental inclusion of fisheries' interests are difficult to separate. Factors apart from legitimacy seemed to be important as well, such as not addressing the right scale of the problem (i.e. not the European scale) and not being able or willing to give financial compensation for damages.

4.5.3 Perceived Legitimacy of the Danish Plan

For the Danish model case, the discourse analysis undertaken within the project gives some insights into the legitimacy perceived by the stakeholders (cp. especially Mathiesen et al. 2004 for this section). It appears from the interviews that all stakeholders share the perception that environmentalists (e.g. ornithologists) had the most influence on the initial and following management plan. Commercial fishers and their national association are particularly dissatisfied with not being taken seriously by decision-makers. Fishers feel victimized by a public perception of nature as something that must be conserved and protected even at the cost of fishermen's livelihoods. Fishers find this expressed in a dominance of cormorant-friendly officials (typically biologists), who are working in public administrations and who are unwilling to involve commercial fishers in the elaboration of the plan. The perception of commercial fishers being shut out from influence is shared by several stakeholders, who believe this is happening because of the lack of scientific evidence supporting the fishers' claims.

The legitimacy of the plan is closely linked to the use of knowledge in management decisions. Managers and several stakeholders depend on the availability of scientific data and documentation of problems. Depending on what research is being carried out and being funded, the stakeholders are given different possibilities to document their problems and hereby influence decision-making.

The Danish case is an example of how important scientific documentation is and how it is linked to interest representation. The conflict has been in a deadlock for years due to the fact that one group of stakeholders lacked scientific documentation for their claims. Very recently, the conflict moved out of this deadlock as new scientific information supporting the fishers' claims became available. As soon as this documentation emerged, stakeholders were put on equal terms and a constructive process was made possible.

4.5.4 Legitimacy in a Multi-level Governance System: The European Cormorant Action Plan

This section sketches out those legitimacy problems encountered by the preparation process of the pan-European Cormorant Action Plan, which are typical for the multi-level governance. The section starts from the assumption that a combination of local, regional, national and EU-wide processes could ideally achieve the best results as such a combination could – at each level – include the relevant scientific and non-scientific information, consider the costs of the process and of the decisions, start positive social dynamics, and gain legitimacy for the process and the decisions. Combining the different levels seems to be a necessary task due to the interdependency of the levels when deciding and implementing an EU-wide cormorant action plan.

Decision-making in a multi-level governance system poses new challenges to questions of legitimacy. The multi-level consideration on **legal compatibilities** mainly focuses on the autonomy each national (or sub-national in federal countries) level demands. Actors on this level use their autonomy to balance internal interests. It is not clear how the existing difficulties to find a common ground with these players can be surmounted in order to build a European-wide cormorant action plan. The member states with the country-respective stakeholders and with the respective weights given to their interests are free to act in negotiating the action plan in the ORNIS Committee. This freedom of action counterbalances an effective management plan calling for a well-defined common aim, specific actions at specific periods and places, an international reimbursement scheme, and a yearly changing set of priorities for the actions. The costs of finding such a solution seem to be judged higher than the gains of moving to a win-win position.

The challenges of **inclusion or representation** of interests depend on the respective level of policy making, but also on the specific situation. This applies to the definition of interests as well as to the factual inclusion of these interests in the process. The institutional setting at the EU level, such as in the ORNIS Committee, allows for participation of NGOs, but the member states definitely dominate the processes – it is their interests, which count. Countries with a tradition of widespread participation, such as Sweden or the Netherlands, deal differently with inclusion and representation than do most new EU member states. These differences are still accentuated at the local level, as we saw in the FRAP project. When working towards an EU-wide action plan, it is not clear beforehand which interests to include; examples could be the interests of the member states, of the local fishermen, or of the national nature protection NGOs.

The **transparency** of the rules and of the process is difficult to maintain. Combining local processes with regional, national, EU and even beyond, in such a way that participants in each process fully understand the way their interactions influence the other processes, seems to be quite impossible. If such processes are to be effective, then they have to look for windows of opportunity, and this on each level. This makes their integration across the different levels a very difficult task.

The European Action Plan, which was never implemented, had no clear **accountability**, even in its preparation. It is still unclear what should have happened with the plan, why it was not implemented, and who could have done something for its implementation. Also, in the ORNIS Committee, only the consensus of the member states can give birth to new actions. Single member states can only demand the elaboration of the plan and the European Commission can only initiate reflection on its necessity.

4.6 Conclusion

Elaborating a trans-national species management plan, not aimed at preserving a species but at mitigating conflicts due to a highly growing population, did not seem to have enough legitimacy. Even a plan that more or less consisted of a list of possible mitigation measures was not agreed upon, as some actors did not believe that the plan respected their interests. National or sub-national plans are judged more or less legitimate, and more or less effective in mitigating local conflicts, but they are also judged as failing to address the increasing dynamics of the problem. Many actors agree on the necessity of a new European-level management plan, but most of them are pessimistic about the possibility of developing such a plan in the next years. The rejection of one main state actor (the Netherlands) and the argumentation supported by scientists that such a plan cannot be implemented effectively seem to dissuade any current effort.

In fact, the effective implementation of a plan aiming at massive population reduction, and only this can reduce the conflict, demands a huge co-ordinated effort, and this particularly from countries that have been rather reluctant to implement population management. Such a plan also demands a sustained effort, as no reason is apparent why the cormorant population should not continue to grow as fast as it did during the last 20 years.

What is rather astonishing, though, is the non-existence of a process elucidating these difficulties, with the aim to elaborate different options apart from massive population reduction. Such options could be highly co-ordinated monitoring of the population, of the killed birds, of the population structure. Two initiatives exist where such a process could be undertaken below the ORNIS Committee level: the cormorant specialist group of Wetlands International, and the already named INTERCAFE project. National authorities are only partly represented in these initiatives, and, until now, the willingness of these initiatives to undertake an effort aimed at a European management plan was not very apparent.

Within the ORNIS Committee, the possibility to draft a plan without all member states adhering to it has apparently not been considered, because there are no possibilities to force a member state to subscribe to such a plan if it did exist. Species management not aimed at enhancing species conservation is institutionally a new problem, and, until now, no structures have been found to deal with this.

References

Bonn Convention (1997). *Action Plan for the Management of the Great Cormorant*. Proceedings of 'Expert Meeting'. Convention on Migratory Species (CMS), Copenhagen.

Carss, D., ed. (2003). *Reducing the conflict between cormorants and fisheries on a pan-European Scale. REDCAFE – Final Report*. Banchory, United Kingdom, Centre for Ecology & Hydrology.

Davies, B. B., K. Blackstock and F. Rauschmayer (2005). '"Recruitment", "composition" and "mandate" Issues in deliberative processes: Should we focus on arguments rather than individuals?' *Environment and Planning C – Government and Policy* 23(4): 599–615.

European Alliance of Anglers (1998). 'Situation of the cormorant in Europe.' EEA, Amersfoort, Netherlands, 48 pp.

European Union (2004). 'Official Journal of the European Union.' Notice No. 2004/C 65 E/170, Subject: Predatory sea birds.

Johnson, H. and G. Wilson (2000). 'Biting the bullet: Civil society, social learning and the transformation of local governance.' *World Development* 28(11): 1891–1906.

Kauffmann, M. (1999). 'Legalität/Legitimität.' In H. J. Sandkühler (ed.): *Enzyklopädie Philosophie*. Hamburg: Felix Meiner, pp. 783–787.

Koed, A., H. Baktoft and B. D. Bak (2005). 'Causes of mortality of Atlantic salmon (*Salmo salar*) and brown trout (*Salmo trutta*) smolts in a restored river and its estuary.' *River Research and Applications* 21: 1–11.

Krämer, L. (2000). *EC Environmental Law*. London: Sweet & Maxwell.

Marion, L. (2003). 'Recent Development of the Breeding and Wintering Population of Great Cormorants *Phalacrocorax carbo* in France – Preliminary Results of the Effects of a Management Plan of the Species.' *Die Vogelwelt* 124(Cormorants: Ecology and Management (supplement)): 35–39.

Mathiesen, C., T. Olesen, D. C. Wilson, R. Varjopuro, O. Zwirner, H. Wittmer, S. Moretti, R. Ferreira dos Santos, L. Vasconcelos, G. Baptista, J. Gomes, L. Madruga, N. Franco, P. Antunes, K. Bruckmeier and C. H. Larsen (ed's) (2004). *Discourse Analysis* (EU R& D project FRAP: Framework for Reconciliation Action Plans). Leipzig: UFZ.

Similä, J., R. Varjopuro, A. Kranz, I. Glitzner, K. Polednikova, L. Polednik, T. Olesen, R. Thum, K. Schwerdtner, I. Ring, L. Madruga, F. Mil Homens, S. Moretti, G. Minarelli, C. Scalabrino, R. Ferreira dos Santos, P. Antunes, K. Bruckmeier and C. H. Larsen (ed's) (2004). *Framework Reconciliation Action Plan: Module Legal and Institutional Basis* (EU R& D project FRAP: Framework for Reconciliation Action Plans). Leipzig: UFZ.

Steelman, T. A. and W. Ascher (1997). 'Public involvement methods in natural resource policy making: Advantages, disadvantages and trade-offs.' *Policy Sciences* 30: 71–90.

Stirling, A. (2006). 'Analysis, participation and power: Justification and closure in participatory multi-criteria analysis.' *Land Use Policy* 23(1): 95–107.

Thum, R. (2004). 'Rechtliche Instrumente zur Lösung von Konflikten zwischen Artenschutz und wirtschaftlicher Nutzung natürlicher Ressourcen durch den Menschen am Beispiel Kormoranschutz and Teichwirtschaft [Legal instruments for resolving conflicts between species conservation and human use of natural ressources, exemplified by the conservation of cormorants and pond aquaculture].' *Natur und Recht* 9: 580–587.

Thum, R. (2005). 'Zur Rechtmäßigkeit so genannter Kormoranverordnungen [The lawfulness of so-called cormorant bye-laws].' *Agrar- und Umweltrecht* 2005(5): 148–152.

Van Eerden, M. (1996). 'Recent meetings about cormorants in Europe.' *WI – CRG Bulletin* 2.

Wittmer, H., F. Rauschmayer and B. Klauer (2006). 'How to select instruments for the resolution of environmental conflicts?' *Land Use Policy* 23(1): 1–9.

Chapter 5
Animal Governance: The Cormorant Case

Hub Zwart

5.1 Introduction: White Trees

While reading Felix Rauschmayer and Vivien Behrens' chapter on the legitimacy of species management a childhood memory came to my mind. One day, as I went fishing with my father, he pointed out to me a number of 'white trees' on the opposite bank of a large shallow fishing pond. The trees were covered with bird excrement and had died. He explained to me that these trees were inhabited by a colony of cormorants: consumers of large quantities of fish, as well as notorious polluters of their immediate environment. I found the scene impressive because, apparently, other species beside humans could have a substantial impact on their environment. We were not the only polluters. This image was activated, so to speak, by reading Rauschmayer and Behrens' chapter on species management. The idea that bird species may be regarded as competitors and polluters is more or less at odds with a more romantic view of nature, which I endorsed when I was young, a view that basically claims that, in a state of nature, things tend to be peaceful and well balanced. We (human beings) are the disruptors.

The views on species management brought forward in Rauschmayer and Behrens' chapter are interesting for a variety reasons. The most obvious significance is, no doubt, that it fleshes out a workable approach to a number of tensions involved in contemporary species management policy on an international level. As a philosopher, however, I find this chapter interesting for another reason. The details of current policy problems in species management are not what I am most interested in. Rather, I find this chapter interesting because it reflects a particular, and highly influential, view on nature.

Hub Zwart
Philosophy, Faculty of Science, Radboud University, Nijmegen, The Netherlands; Department of Philosophy & Science Studies (Chair); Centre for Society & Genomics (Director); Institute for Science, Innovation & Society (Director)
E-mail: h.zwart@science.ru.nl

5.2 Views on Nature

Roughly speaking, three basic views on nature can be distinguished. First of all, there is the view that nature is basically a wilderness. Species as well as individual organisms find themselves in a permanent state of war. Human society is basically an artefact, a post-natural enclave. Within society we try to manage our affairs in a fair, peaceful and reasonable way, but beyond the confines of the human realm, a continuing struggle for existence remains the fate of all species – all but one. And should we no longer actively commit ourselves to establishing and maintaining a humane and rational society, we will sooner or later sink back into a 'state of nature' – a dreary prospect.

An opposite view sees nature basically as a realm of symbiosis, equilibrium and harmony. This is what I already referred to above as the romantic view. We humans are the disruptive factor. We ourselves are responsible for disturbing the natural balance. In principle, human societies should also reflect this balance of nature. Instances of societal tension and crisis are interpreted as symptoms of estrangement and degradation with respect to a more natural state of affairs.

The chapter by Rauschmayer and Behrens reflects a third view, a much more contemporary one. We have come to look upon both views outlined above (the 'rationalistic' and the 'romantic' view) as highly ideological. Our own current views on nature claim to be less ideologically framed – at least that is what we ourselves believe. Current views claim to be more or less evidence-based, and to a certain extent this is true. Although rationalism and romanticism have inspired impressive research programs,[1] the broad and comprehensive views that constituted the core of these perspectives are now regarded as ideological and normative. But also our own current views of nature, such as articulated in their chapter, reflect the way we (would like to) see ourselves, the way we (would like to) see our own society. We intend to manage nature (and conflicts in nature) in a way that is reminiscent of how we manage ourselves, our own conflicts. In this new vision of nature, which I will refer to as the *governance view* on nature, various species (including cormorants and humans) are seen as stakeholders, and we somehow must find ways to manage and regulate their various interests and claims. We must monitor the behaviour and adjust our policies and regulations to our findings and evaluations. Other species beside human beings have an impact on their environment, a 'footprint': they consume scarce resources and pollute their surroundings. An institutional framework has to be developed that allows us to address and manage possible inter-species tensions and conflicts.

This calls for a more or less procedural form of inter-species justice and legitimacy. The relevant interests of stakeholders (species) have to be identified and represented. Forms of procedural fairness have to be put in place.

---

[1] The work of Alexander von Humboldt for example, although of great scientific significance, was clearly inspired by a particular view of nature, namely the romantic one: nature as a comprehensive landscape, a harmonious whole, both scientifically and aesthetically appealing. See for instance: Alexander von Humboldt (1808) Ansichten der Natur. Stuttgart/Tübingen. [Views of nature. New York: Arno Press, 1975].

The romantic view stated that in principle, active interference with nature is bad, unless its objective is to re-establish a viable situation we ourselves had disrupted previously. Its starting point was an a priori predisposition against interference. Nature has its own resources and talents for self-management. We must interact with nature in such a way that we use and respect the built-in dynamics of natural systems. Allow nature to do her work: be patient. Do not underestimate nature's potential. This view implies an element of trust in nature, a willingness to rely on nature's own resources for self-governance. The romantic view clearly comes into conflict with the rationalistic view on nature, where nature is basically seen in a much more negative light, as the realm of the non-human, the non-moral, the non-rational.

The governance view on nature is guided by the idea that nature can be brought into our regulatory frameworks and institutions. This more or less novel approach to nature has a profile of its own. Although to a considerable extent it can be regarded as research-based, the research questions and research findings it engenders are framed by this normative guiding idea: that nature is neither a wilderness, nor an equilibrium, but rather an arena where interests and claims of various stakeholders (species, including our own) may come into conflict.

5.3 The Governance of Nature

The chapter by Rauschmayer and Behrens articulates (and builds on) this view. It intends to show the legal and institutional background of nature and especially bird conservation as well as efforts that have been undertaken to agree on a European management plan. This framing of the chapter's basic objectives already outlines the basic stance that is taken towards nature. We are responsible for nature. Nature cannot be expected to care for itself, so to speak. Without governance, cormorants would loose their struggle for existence, or harm the interests of other stakeholders. Our responsibility is not defined in terms of a top–down perspective, as would have been the case if a rationalistic perspective had been adopted. We cannot simply impose our rational laws. Rather we must enter into a complicated process of negotiation and deliberation, of monitoring and evaluating, in which the claims and interests of both cormorants and humans must be taken into consideration. We learn that a cormorant eats approximately 0.5 kg fish per day, and that this re-introduced species therefore represents a new competitor in the fishery arena, endangering the interests of professional and sport fishermen. A 'cormorant management plan' is therefore called for which can help to secure the fisheries' interests while preserving 'at least the minimum viable population of the target species' (3). The cormorant does not simply live 'out there'. Rather, this species inhabits a world that is already densely structures by Directives and Conventions. By intricate and research-based policy procedures, this landscape is to be modified in such a way that tensions can be reduced so that they become manageable, while interests can be balanced. The symbolical elements of the governance landscape are no less important, when it comes to determining the prospects of a species, than the natural ones. Indeed, it has become difficult to think about natural elements outside a governance frame. This frame determines, for example, whether the cormorant can be regarded an endangered species, in what areas it is allowed to dwell, and whether and under what conditions 'derogation rules' apply.

5.4 Constructing Legitimacy

The chapter describes a series of events in which the governance view on nature comes to live, so to speak, namely a series of meetings (Lelystad, Copenhagen, Geneva) where available information is collected, technical and financial measured are considered and monitoring strategies are planned. Participants discuss issues such as the optimal number of cormorants for viable conflict management and the legitimacy of cormorant reduction in the light of international and EU legislation. Various stakeholders are represented, such as fishery organizations, scientists and species conservationists. It is pointed out that, due to recent changes, like changes in human behaviour (increase of sports fishery) or in animal behaviour (increase of migrations, dissemination of novel hunting techniques), inter-species conflicts may escalate. Not the details of these deliberations are of interest here, but rather the general framing of the process, the key concepts the participants adhere to, the style of reasoning that is at work. Typical elements of a governance perspective such as information management, decision process and implementation strategy are duly addressed. Measures are perceived as legitimate if they are in accordance with the prevailing rules and procedures of decision-making. Information produced by monitoring strategies is an important input to the deliberation process. In order for a set of measures to be legitimate, the mere fact that they are generated by the legitimate authorities is not enough. They also have to be *perceived* as viable and legitimate by (representatives of) the various stakeholders involved. This calls for an interactive procedure, an interactive form of policy development. In other words, stakeholders (or their representatives) have to become active participants in the process. Whether or not a particular policy is regarded as legitimate may depend on various factors, mentioned in Rauschmayer and Behrens' chapter, such as the question 'whether all relevant interests are represented adequately and whether the procedure permits protection or, ideally, enhancement of the interests involved' (p. 68). Other important issues have to do with transparency and inclusion/exclusion as well as the question whether and how the least powerful stakeholders are involved in the process.

An important issue in such a context pertains to the availability a various forms of knowledge. Expert knowledge of environmental scientists for example, and legal knowledge of policy experts, has to compete more or less with practical and experiential knowledge of individuals involved in fishery. In a governance perspective, no single form of knowledge can be regarded as overriding and 'objective'. Rather, all knowledge claims emerge from a particular perspective, they all have their strengths and weaknesses and they all have to be given due consideration in the deliberation process.

Meanwhile, the cormorants themselves live in a world of their own where the governance dimension will be completely absent. They have their own forms of 'knowledge', their own sources of information, inaccessible to us. Governance refers to the way in which their world is represented and reconstructed *by us*. We see the activities of cormorant colonies in this light. It is the clearing that allows their world to become visible to us in a certain way. It is an anthropogenic perspective of fairly recent origin. For us as contemporaries, however, it becomes difficult, if not impossible, to see their world in different terms. Alternative views, such as the romantic one, have lost much of their former credibility. The governance view on nature has become an a priori format, an ethical paradigm that is difficult to ignore. This gives it the aura of a more or less impartial view. Increasingly, debated on species management will be stylized in conformity with this prevailing 'thought style'.[2]

---

[2] Ludwig Fleck (1935/1979) Genesis and development of a scientific fact. Chicago & London: University of Chicago Press.

Part III
Protected Areas

Nature Park Donana (Spain) (by Susana Aguilar Fernández)

Chapter 6
Legitimacy Problems in Spanish Nature Policy

The Case of Doñana

Susana Aguilar Fernández

6.1 Introduction

As far as European biodiversity is concerned, Doñana is undoubtedly one of the 'jewels in the crown'. Located in the south of Spain (Andalusia) and covering over 117,000 hectares, it contains three main types of ecosystems (wetlands, mobile dunes and Mediterranean forests) and hosts over 875 plant species, 226 types of birds (among them, the almost extinct imperial eagle), important fish, and reptile and mammal life (the lynx being its most significant example). In particular, Doñana is a winter resort for more than 6,000,000 migratory birds. As a result of its unique natural assets, Doñana was declared a Wetland of International Importance by the Ramsar Treaty, as well as a Biosphere Reserve (1980) and a Patrimony of Humankind (1994) by UNESCO. Under the EU Birds Directive, Doñana became a special protection area. It has also received the European Diploma of Protected Areas, and has been integrated into the Natura 2000 European network.

The recognition of Doñana's importance as a natural site can be traced back to the fifteenth century, when certain limits on settlement in the area were imposed in order to protect its rich hunting resources. Two centuries later, the Duke of Medina Sidonia ordered the building of a palace for his wife, Doña Ana Gómez de Mendoza, at the very core of the natural site. Places in the neighborhood began to be known as The Forest of Doña Ana, The Hunting Reserve of Doña Ana and so on until the name acquired its current elided form: Doñana (Figueroa and Toro 2003).

Susana Aguilar Fernández
Complutense University of Madrid (Spain)
E-mail: saguilar@cps.ucm.es

J. Keulartz and G. Leistra (eds.), *Legitimacy in European Nature Conservation Policy:*
Case Studies in Multilevel Governance, 83–100. © Springer 2008

6.2 Initial Conservationist Impetus: 'Enlightened Despotism' or the Lack of Input Legitimacy

Scientific interest in Doñana gradually began to emerge in the nineteenth century after the publication of a catalogue of birds in Andalusia by a well-known conservationist, Antonio Machado. However, the landowners of Doñana continued their hunting activities while introducing new plant species that put the ecosystem's equilibrium in peril. Around 1952, ornithologists from all over the world put forward the idea that Doñana should become a scientific reserve. Growing conservationist awareness led to the acquisition of nearly 7,000 hectares in the area by the World Wildlife Fund (WWF) in 1963. This group then proceeded to hand the property over to a scientific institution integrated into the Spanish Ministry of Education, the CSIC. The area was subsequently declared a biological reserve and in 1964 the Biological Station of Doñana was set up. The biologists who created this station would later become crucial in two ways: as the vanguard of a new conservationist mood that would clash with the production-oriented approach of the main state agency in charge of conservation (the Institute for the Conservation of Nature, Icona), and as a trigger for the creation of ecological associations in Spain, specifically in the Andalusian region (Fernández Reyes 1993).

In 1969, the government of Spain set up the Doñana National Park,[1] and in 1978 its surface area was enlarged by legislation (Law 91/1978). The law not only increased the size of the national park, but it also gave Icona new powers over the management of the area, while underlining the research functions of the Biological Station. The upgrading of the tradition-oriented Icona to a managing unit helps explain the tense relationship between this state agency and the forward-looking Biological Station. Likewise, the law envisaged a decision-making body (Patronato) that would allow public authorities (from the national, regional and local levels), scientists (the directors of the Biological Station) and interest groups (environmentalists and farmers, among others) to come together. Finally, it introduced a territorial coordinating instrument (*Plan Director Territorial de Coordinación*, PDTC) to prevent the production activities taking place in the surroundings from impinging upon the national park (Van der Zouwen 2006). In a sense, the PDTC anticipated the concept of sustainable development (SD) (although the expression does not appear as such in the text), when articulating the aspiration that Doñana conservation would be compatible with the human-related activities (basically, farming and building) occurring near its boundaries. The approval of a management plan (*Plan Rector de Uso y Gestión*, PRUG) for Doñana suffered an enormous delay in the meantime; it

---

[1] The definition of a national park is that of a natural area, little transformed by human settlement or production activities, that, on the grounds of its landscape beauty, ecosystem significance or the uniqueness of its flora, fauna, or geomorphological formations, possesses ecological, aesthetic, educational or scientific values whose conservation deserves preferential treatment. A national park, therefore, is considered to be of general interest for the nation, its declaration adopting the form of a statute.

was only in the year 2004 that the national park could avail itself of this essential instrument, *reglament 48/2004*.

Under radically different political conditions (a democratic and decentralized state being a reality), the regional government of Andalusia established the Doñana Nature Park in 1989, which gave further protection to the area surrounding most of the then existing national park. At present, the national park, whose management falls jointly under the national and the regional government, covers 64,260 hectares and embraces parts of the provinces of Huelva and Seville, while the nature park, which is exclusively a regional responsibility, stretches over 53,709 hectares and also includes part of the province of Cádiz.

It is important to bear in mind that the first initiatives to protect Doñana were clearly top–down, embracing an odd alliance of national and international conservationist figures with some prominent officials, who were gradually taking ecological issues on board. The preparations leading to the United Nations-sponsored Stockholm Conference on Development and the Environment in 1972 were the main factor accounting for the emergence of a new ecological interest in some minority sections of the administration. This interest was otherwise absent in an authoritarian regime that had embarked on the pursuit of unbridled economic growth since the late 1950's (Aguilar Fernández 2004). As a result, lack of both consultation and participation of interested parties can mainly be accounted for by the non-democratic political nature of the Francoist regime. Leaving aside the fact that collective decisions leading to the conservation of Doñana did not take into account the wishes of locals (non-existing input-oriented legitimacy), it is undeniable that actions related to the preservation of the site were urgently needed, if its visible ecological deterioration was to be stopped before it became too late to conserve its unique natural assets.

That urgency in nature policy can sometimes lead to 'ecological dictatorship' has been raised by some authors, and that this lack of input legitimacy is frequently associated with the recurrent emergence of social problems and citizens' resistance to implementing nature management plans has been frequently discussed also.[2] If democratic regimes sometimes cannot escape this two-fold problem, the probability that nature policy decisions under non-democratic conditions become socially contested, encountering serious boycott strategies on the part of the interested parties at the implementation stage, is much higher. In a sense, the difficulties that have constantly flanked the management of Doñana might be explained by the peculiar political and social circumstances (a combination of political authoritarianism plus top–down conservationism) under which the area was initially preserved. More important today though is the ambitious, but weak participatory conservation agenda that was adopted by the Andalusian regional government in the late 1980's,[3] once the decentralization process initiated

---

[2] Social participation must be seen as a necessary (but not sufficient) condition for the success of nature policy. As stressed by Meadowcroft (2004), if target groups and citizens are not brought into the policy-making process, enforcement problems will clearly arise.

[3] An ambitious nature policy accompanied by little social participation under democratic circumstances can be labelled, following Easton's formula, as 'authoritative allocation of values' regarding conservation.

after the approval of the 1978 democratic Constitution was in full swing, and that has been in effect ever since. Since regional governance in nature policy (output-oriented legitimacy) has frequently been denounced by environmentalists and experts alike for paying lip service to participatory procedures, decisions concerning the conservation of Doñana have been generally disputed. The same argument, but with purposes other than conservation, has been used by locals when reacting against different protection schemes that would allegedly respond to pressures exerted by outsiders not living in the area. Last but not least, the powerful pro-development interests that operate in the area obviously need to be taken into account when coming to terms with Doñana's 'troublesome environment'. These interests have encouraged alliances with local groups (municipal authorities, farmers and cattle-breeders), calling into question the legitimacy of decisions on nature policy that, characterized as non-participatory and non-transparent, are detrimental to the economic well-being of the neighbors. Collective action with the aim of hampering the effectiveness of government policies in relation to Doñana (thus contributing to flawed output-oriented legitimacy)[4] has been one of the most visible results of this alliance. The need for production and conservation interests to engage with one another, as expressed in the SD strategy worked out for Doñana in the 1990's, has not substantially contributed to reducing the tensions in the area.

6.3 Spanish Nature Policy: The Slow Convergence with Europe

Spain was one of the pioneers among European countries in introducing a conservation policy, designating its first two national parks in 1918: Covadonga (later on, Picos de Europa), and Ordesa and Monte Perdido. At the start, those areas selected as national parks were of a markedly mountainous nature, Doñana was the first relevant example of a national park that did not share this morphological feature.

An early beginning, however, does not imply that conservation policy was whole-heartedly pursued. A closely linked policy area, forest policy, was clearly production-oriented and was given priority over conservation. Traditionally, forest management was conceived of as a means of supplying the timber industry with raw material. Its predominantly short-term approach (based on the planting of quick growing species of pines and environmentally-damaging eucalyptus) was, however, at odds with the non-productive long-term perspective advocated by the minority groups who took conservation seriously (Elías Mortera 1999). A further proof of erratic nature policy was that many protected sites lacked management plans for a long time, while conservation agencies were deprived of resources and qualified personnel. A narrow-minded approach towards conservation was one of the characteristics of this old-fashioned policy.

New legislation in 1975 (Law 15/75) and its regulations passed in 1977 (*reglament* RD 2672/77), paved the way for a new nature policy during Spain's democratic transition. This law 'not only laid the foundations for a legal regime for

---

[4] A discussion of the three dimensions of legitimacy can be found in Scharpf (1998).

Doñana national park, but also clearly drew the boundaries between the tasks and responsibilities of Icona and the Biological Station' (Van der Zouwen 2006:107). Based on this law, Icona started to catalogue natural sites demanding special protection. From an initial estimate of 500,000 hectares, the area to be potentially preserved increased to over 3,000,000 hectares – 6% of the country's surface – and, in 1980, this was further increased to around 10%. The first draft catalogues basically covered publicly owned woodlands (87%), wetlands and other types of natural sites. Public interest in a broader range of natural sites, as shown by the state agency when cataloguing wetlands (a natural asset that had been systematically neglected), was in line with both European policy and some new all-encompassing concepts in ecology science.[5] Yet a novel approach in nature policy that is more centred on biologists' expertise and holistic solutions has only existed substantially since the 1980's. Thus, the traditional pro-development stance of Icona gradually started to change only after Spain's accession to the EU. Behind this change two important factors can be identified: 'On the one hand, the entry of biologists in Icona...reinforced the criticisms of the conservationist movement about the work of forest engineers as the majority staff at Icona; on the other hand, the person who then occupied the position of Icona director (J. Lara) was clearly in favour of a reorientation of conservation policy' (Aguilar Fernández and Jiménez 1999: 237).

When Spain began its democratic transition in 1977, the administration for environmental protection was characterized by a marked fragmentation of powers. This problem was partially compensated for by the gradual upgrading of the central administration,[6] which finally led to the creation of the Ministry of the Environment (MIMAM) in 1996 after the right-wing Popular Party (PP) acceded to power, following fourteen years of socialist rule (1982–1996). Once established, MIMAM not only ended the 'exceptionalism' that had made Spain the only EU country lacking a department of this type, but it also furthered the concentration of environmental powers because now, for the first time in Spain's history, conservation and environmental powers fell under the remit of the same department (Aguilar Fernández 1997b).

---

[5] 'Since the 70's, and especially since the late 80's, nature policy has...been developed into an independent policy sector in many European countries...This concept (of ecological networks) has increasingly dominated nature policy since the late 80's. It runs parallel with the rise of the notion of "ecosystems"...Other concepts also have come in vogue over the last decade. "Sustainable development" and "biological diversity", or "biodiversity" are examples that have found their place on the political agenda.' (Van der Zouwen 2006: 2–3).

[6] In 1977 responsibility for the environment was transferred from the Presidency of the Government to the Ministry of Public Works (MOPU). One year later, MOPU created the General Directorate of the Environment, which was replaced by the General Department of the Environment in 1990 (SGMA). As a result of the transformation of MOPU into the MOPT (Ministry of Public Works and Transport) in 1991, the State Department for Water Policies and the Environment took the place of the SGMA. After the 1993 elections, this unit was renamed the State Department for Environment and Housing, and the MOPT added the term 'environment' to its name, becoming the MOPTMA.

Before 1996, the Ministry of Agriculture (MAPA), or, more precisely, Icona, had been the unit responsible for conservation policy. MAPA never took conservation very seriously and left Icona a free hand to make policy. Icona, created in 1972, was disbanded in 1995 once its presence had become untenable for the regions, which insistently claimed that the agency interfered with their powers. Shortly before the disappearance of Icona important changes took place within MAPA, which helped to patch up its fragmented nature by concentrating conservation powers in a new institution, the National Commission for the Protection of Nature (CNPN). The CNPN was created by Law 4/89 for the Conservation of Natural Areas and Flora and Fauna and regulated by the *reglament* RD 2488/94 with the aim of coordinating conservation policy between the state and the regions.[7] In 1995, one year before MIMAM took over conservation powers from MAPA, an autonomous agency for national parks was set up. Law 4/89 was subsequently modified by Law 41/97, which, together with the *reglament* RD 1760/98, further changed conservation policy by establishing a council for the network of national parks (with non-binding powers) and a mixed-management commission, which allowed national and regional authorities to discuss strategies for these parks on an individual basis. The modification of Law 4/89 was brought about by a Constitutional Court decision that largely sided with the regions when they complained that exclusive state responsibility for national parks was unconstitutional. The plea for joint management in these areas was thus granted by the court and took the shape of the above mentioned mixed-management commissions. This is the reason why the national park of Doñana is, at present, the responsibility of both the national and the Andalusian governments.

In essence, a new environmentally sensitive approach to nature policy along with the upgrading of the administration responsible for this policy has entailed some sort of convergence with European developments (Aguilar Fernández 1997a). This is not to say that Spanish conservation policy has been totally 'Europeanized' or that the long endogenous tradition of nature management has been wholly replaced by exogenous practices. As a matter of fact, Europeanisation of Spanish policy and domestication of European policy have gone hand in hand.[8] A proof of 'the European imprint' has manifested itself through the Birds Directive, which has led to 148 areas of special protection for birds (ASPB's) being declared, covering approximately 2,500,000 hectares. Europeanisation was also clear in Law 4/89 for the Conservation of Natural Areas and Flora and Fauna and the *reglament* 97/95, which, referring to the Natura 2000 network, made the state responsible for putting forward ASPB's in Brussels, while regional governments were entrusted with their management. On the other hand, 'the efforts to domesticate EU policy have been mainly

---

[7] The CNPN used to function through two types of forum: one in which general directors were brought together and another in which experts of the administration gathered in four different technical committees (natural areas, flora and fauna, wetlands, and forest fires).

[8] A discussion of the concepts of Europeanization and domestication can be found in Jordan and Liefferink (2004). In short, Europeanization is 'a top down process impacting on states' (6), while domestication is a bottom up process stemming from specific Community countries and impacting on the EU.

deployed in two fields: water policy and conservation policy. . .Conservation issues have been paid great attention because the costs associated with the maintenance of some of the richest biodiversity in Europe had allegedly not been duly considered by the EU when it designed its habitats policy' (Aguilar Fernández 2004: 175–7). A brief account of the Spanish strategy in relation to the Habitats Directive will illustrate this point.[9]

The Habitats Directive was negotiated for a full five years (1988–92) because, among other reasons, the Spanish government rejected it on the grounds that Community financial instruments for conservation were missing. In order to ensure Spanish interests in Brussels, the director of Icona gave the Spanish delegation clear instructions relating to the financing mechanisms of the directive. Firstly, the delegation was to make certain that no 'disproportionate' expenses should ever be imposed on the members states, and secondly, that if expenses were unavoidable, the Spanish delegation should then proceed to defend the need for some European economic contribution. Being a controversial issue heavily disputed by some European 'heavyweights', such as Germany and the United Kingdom, the need to introduce some Community economic support was not taken on board in Brussels until the end of 1991. Only when the Spanish delegation publicly affirmed that Spain would not spare any effort to protect habitats provided that Community money was used to balance an unbalanced situation, did the Commission begin to grasp the full economic implications of the directive. Subsequently, a Spanish ornithological group (SEO/Bird Life) drafted a document, which, after being leaked to all member states, helped decrease the European Council's reluctance to co-finance the directive. As Spain continued to block the text, claiming that the extent of financial support should be established and that the reception of funds could not be considered 'an exceptional case' (Quercus 1991), the SEO prepared a second proposal, which fixed the level of European money to be given to the less developed regions with habitats of priority interest. Only when it became clear to the SEO that the Spanish government was using the habitat issue as a legal precedent to achieve direct Community financing for future directives, did it decide to send the text to all the member states, the Commission, and the Dutch Presidency of the EU. The final obstacles to passing the directive were eventually removed after concerted European pressure was exerted on the Spanish government.[10]

The Spanish input in the text was of the utmost importance, as it managed to alter the initial central-northern bias of the directive so that it became more sympathetic

---

[9] This account is based on Aguilar Fernández (2003).

[10] In 1992, the Habitats Directive (92/43) was approved in Brussels and translated into national law in 1995. The threatening position of the Commission, which repeatedly accused Spain (as well as other member states) of infringing on the directive because of its inability to meet the deadline for its translation into national law (June 1995), helps to account for early contacts between national and regional authorities under the auspices of Icona, first, and the CNPN later on. By the end of 1995, a number of committees (one of them related to protected spaces, the Committee of Wetlands and Natural Species) were created at the CNPN. Two main issues under discussion were the elaboration of habitat lists and the expenditure estimates.

to the ecological situation of southern Europe. In this sense, Spain took the lead in an effort to bring Mediterranean environmental problems into the picture. Not only did the directive eventually mention European co-financial contribution in its 8th article, but the southern perspective also pervaded the entire text. Thus, Article 6 of the directive distinguishes between protection (which entails a restriction of activities in the protected areas) and conservation (associated with the idea of sustainability, or the compatibility between environmental protection and economic development); refers, in its introduction, to the need to count on financial support in order to implement the directive; accepts a laxer application of the directive, in terms of selection criteria and financing mechanisms, if the country protects more than 5% of its territory; and, questioning the meaning of priority habitats, restricts the protection of certain species (such as the lynx and the brown bear) to smaller areas.[11]

6.4 Decentralization or Hollowing Out the State's Role in Nature Policy

Parallel to European convergence, a process of decentralization, through which the regions have become crucial political actors, has been gathering momentum in Spain since the beginning of the 1980's. The problem of regional reinforcement has been that it was not linked to an institutional framework designed to coordinate the increasing number of administrative units with nature policy responsibilities. The accession of Spain to the EU highlighted this deficient coordination because of the need to fulfill a growing number of highly technical Community requirements. The establishment of new regional environmental departments has not improved the poor Spanish environmental record, while the drawbacks associated with troublesome intergovernmental co-ordination have undoubtedly contributed to the high number of breaches of Community environmental rules. Fragmentation of responsibilities and insufficient coordination have repeatedly been cited by the European Commission as the main causes of Spain's failure to meet the EU's environmental policy targets. Obviously, the entry of Spain into the EU in 1986 facilitated the approval of several projects whose aim was to fight deficient coordination, both on the

---

[11] Spanish success was based mainly on technically sound reports, which drew the Commission's attention to the drawbacks and inaccuracies of different directive drafts, and on the use of a legitimizing pro-EU discourse. This last point has to be placed in the more general context of a socialist national government which, not having to rule in coalition and enjoying a relatively free hand in Community affairs, was willing to be seen as a main defender of the European cause. In this sense, the Spanish government never disputed the right of Europe to legislate on natural areas nor did it ask for the 'nationalization' of certain powers, which had been gradually acquired by the Commission in relation to this policy (as was the case at the time of the Thatcher government in the UK). The Spanish government only asked for the recognition of Southern ecological specificities while highlighting the fact that conservation in the region was going to be comparatively more expensive and needed, therefore, some EU help.

traditional horizontal axis (different ministries in the national administration) and on the new vertical one (the state and the regions). At present, horizontal coordination has been helped by the creation of MIMAM, not only because the probability of boundary conflicts has diminished but also because conservation powers and water administration now fall under its remit. Yet, the dispersion of environmental responsibilities at the sub-national level (or vertical un-coordination) has been left largely untouched by the upgrading of the national administration (Aguilar Fernández 1997b).

The way the decentralization process has been evolving in Spain seems to be running counter to recent trends affecting nature policy in federal states. Growing emphasis on shared powers along the lines of cooperative federalism, as advocated by some European countries, has been systematically disputed by many Spanish regional governments and not simply the 'historical autonomies' of Catalonia and the Basque Country. On the whole, regions have not recognized (or have even ignored) the state's responsibilities in conservation issues, while the central administration has not played an active role for fear of being accused of interference. Neglect of, if not hostility to, the national government's role in nature policy is reflected in the region's use of the Constitutional Court as a mechanism of defence against 'the state's meddling'. In the end, the Spanish state has become a minimal one whose only powers are external representation and coordination (and even coordinating committees do not work to their full potential because some regions, preferring bilateral mechanisms, object to their functioning, do not attend their meetings, or refuse to send the required information). More precisely, after the transfer of conservation powers to the regions, MAPA was basically left with only two responsibilities: the management of the national parks network and the intergovernmental coordination of the implementation of EU directives. Moreover, the first responsibility was further limited after several regions, including Andalusia, Catalonia and the Basque Country contested Law 4/1989 before the CC, claiming that it interfered with their conservation powers. In 1995, the court ruled that the running of national parks was not the exclusive task of the Spanish state but, at the same time, reinforced the coordinating responsibilities of the central administration. The management of national parks, one of the few powers still in the hands of the national government, was then severely curtailed. This action was received by environmental groups with dismay in the face of recent attempts by some regional governments to decrease the degree of protection in those areas.

The main problem with nature policy, however, is not the absence of shared powers between the state and the regions, or even the lack of a broad, coherent and long-term strategy in conservation policy, but the fact that, more often than not, those regional governments that defend their nature powers so jealously are the same ones that prove to be reluctant to exercise them when it comes to planning policies.[12] The relationship between nature and planning policies is obvious. The

---

[12] In the case of the regional government of Valencia, for instance, its 'permissive attitude...(has meant that it) is not complying with its duties related to the elaboration and coordination of

mismatch between the two becomes apparent when aggressive and environmentally unsound planning schemes destroy ecologically important but unprotected areas or put the surroundings of the preserved sites in peril. These problems contribute to the consolidation of an unsustainable map of 'islands of protection', as the case of Doñana will later show.

In a context in which municipal governments are afflicted by chronic debt, local incumbents have approved, or are in the process of approving, irrational, chaotic and unsustainable planning schemes. These schemes either take the approval of regional governments for granted (the cases of Valencia[13] and Murcia) or do not seem to be encountering enough opposition from them to deter them (the case of Andalusia).[14] At the local level, deeply entangled building and political interests are bringing about the destruction of the Spanish coast. Agency takeover and corruption are most visible in this case. Worse still, the areas that have been afflicted with 'building fever' have recurrent water shortages. Ironically, these are also the areas that have the highest number of golf courses. In the meantime, many locals, blinded by the perspective of short-term economic benefits and easily manipulated by demagogues who talk about comparative grievances, seem to have forgotten that Spain was recently engulfed in a 'war for water' that pitted north-eastern regions (the donors) against south-eastern regions (the recipients). That this war was mainly fought in the name of the right of the less developed regions to develop their farming practices, and that the final result was the approval by the socialist government, which came to power in 2003, of a New Water Plan that did not envisage water transfers but the building of desalination plants in those areas in need of water, is not taken into account either. That the EU, under enormous pressure from the

---

inter-local Territorial Plans' (Romero 2005). Besides, indiscriminate building schemes in the countryside, which are not reflected in the municipal Territorial Plans have also become a general practice. Consequently, regional General Plans tend to be neglected by local authorities.

[13] The only element of the general planning scheme approved by the Valencian regional government that the EU has severely criticised and the only one in which it seems to have some capacity to intervene, is procurement rules, that is to say, the conditions governing competition between private building companies in search of public contracts and licenses, which have been denounced as being biased and not transparent enough. Other elements of the planning scheme, such as the new figure of the planning agent, which, in combination with the Land Law approved by the previous right-wing government, further liberalises access to land, on the irrational assumption that leaving this asset in private hands will contribute to decreasing property prices, will probably not come under attack from the EU. Almost certainly, the EU will raise its voice against a programme that breaches environmental legislation and clashes with the SD agenda. However, European indignation will not have legal implications because its impact will be confined to extensive media coverage. In the meantime, more than a million houses will be built at the seaside, and the domestic courts in charge of overseeing the planning law will not be able to stop the process in time.

[14] According to Romero, the planning of Mediterranean coastal areas by the Andalusian government has completely failed. More than ten years have passed since the approval of the regional Planning Law (*Ley de Ordenación del Territorio*) and the government 'has not been willing or has been unable to give full content to the sub-regional planning schemes'. Such a failure cannot easily be explained since the CC has frequently stated that, under the coordination principle, regional governments are entitled to undertake territorial responsibilities at a multi-municipal level.

different contending parties, finally recommended a sustainable use of the scarce resource, implicitly accepting the thesis of the government, has also been duly ignored.[15]

6.5 The Ambitious Nature Policy of the Andalusian Government: Doñana in the Spotlight

In Andalusia the decentralization process began in 1978 with the approval of Law RD-L 11/78 that set up the embryo of the autonomous community. The establishment of the new regional government roughly stretched up to the year 1982. Once the regional constitution (*estatuto de autonomía*) was passed in that year, the first regional elections were announced. Since then, the socialist party (PSOE) has been in power.

Initially, responsibility for the environment fell upon officials who, in most cases, shared a strong environmental awareness. Transfer of conservation powers from the national to the regional government took place in June 1984, meaning that regional authorities did not enjoy real capacity to carry out any type of nature policy until the mid-1980's. This is the main reason why they initially focused their attention on the identification of urgent problems and on establishing the foundations of nature policy. Until 1984, therefore, declaration of protected areas still fell under the remit of Icona, the agency that previously had to take into account the reports of the Interministerial Commission of the Environment (site of the Committee in Defence of Nature and the Environment). In spite of the lack of conservation powers, some key conservation-minded figures within the regional government took advantage of the existence of the General Direction of Planning (whose powers had been devolved in 1981) to elaborate provincial protection plans. This initial stage, which lasted approximately until the mid-1980's, marked the 'golden age of regional conservationism', visible in a well-established alliance between politicians, civil servants, and experts of the environmental movement, and accompanied by a strong participatory ethos (as manifested by the creation of the Advisory Council on the Environment). Gradually, Andalusian nature policy began to gather momentum as it departed from the state's traditional focus on preservation of woodlands and adopted a broader ecological perspective that showed itself in the Natural Sites Catalogue and the Law of Andalusian Natural Areas, both from 1989.[16] Even more important than this

---

[15] According to a recently published international environmental index, Spain is located near the bottom of the environmental ranking of developed countries (20th position out of 29). Among Spain's main environmental problems are excessive water consumption and deficient nature conservation. According to the director of the study, the Yale Professor Daniel Esty, Spain is growing economically to the detriment of the environment and depends heavily on the expansion of the building sector (*El País*, 24 January 2006).

[16] At the end of the 1970's the deterioration of Andalusian wetlands became the main concern of one of the most important conservationist groups in the region, Andalus. Thus, an important change in the policy discourse, or in social perception, regarding wetlands took place during the

broader focus was the fact that regional authorities perceived the need to provide the emerging Andalusian government with some political legitimacy. This led the authorities to emphasize the difference between the new regional nature policy and the old national one, through active protection of Andalusian biodiversity. This promoted social pride in the unspoiled and rich landscape of the region. As a matter of fact, during the creation of the Andalusian government, regional authorities fiercely opposed several projects promoted by the state. Had these projects been approved, they would have caused the severe deterioration of some important natural assets, the most significant example of which is the Odiel wetlands (Aguilar Fernández and Jiménez 1999). In 1984, the year in which conservation powers were finally transferred to the regional authorities, the Environmental Agency was created and eventually integrated into the Andalusian Department of the Environment, set up in 1993. The establishment of this department entailed the culmination of an upgrading process affecting the regional administration responsible for conservation.

In the late 1980's, the initial emphasis on participation that had permeated Andalusian nature policy,[17] and that had shown itself, for instance, in the detailed and participatory elaboration of specific laws for every site to be protected, began to lose steam. 'Although the elaboration of the (natural sites) catalogue was accompanied by some negotiation with the municipalities and conservationist associations, a new trend, whereby the openness of the nature policy style in its initial stages (was being) reduced, became evident' (Aguilar Fernández and Jiménez 1999: 227). Growing reluctance to promote social participation might have had to do with conflicts relating to the approval of management plans (PRUG's) for natural sites such as Cazorla and Grazalema. That preservation tended to become conflict-ridden when management plans spelt out the rights and duties of the interested parties is undeniable. Both the lack of experience of regional environmental units in dealing with these conflicts and the concentration of administrative resources on the drawing up of the catalogue of natural sites might have aggravated the situation. Finally, the content of the PRUG's became less precise and more vague during the 1990's as a way of offsetting social opposition. Less attention on detailed elaboration was accompanied by a new political mood less keen on opening channels for social participation.[18] In this new context, the regional initiative for Doñana took place.

---

following decade. From unhealthy and ecologically-uninteresting areas, wetlands became an important example of biological diversity. Furthermore, the presence of biologists in the Biological Station of Doñana contributed to this new understanding.

[17] Social participation in nature policy was also present at the state level. As a matter of fact, the transition to democracy was accompanied by an open political style in relatively new and not very salient policy areas, such as that of conservation. This openness was reflected in the incorporation of experts and members of the conservationist movement into the administration.

[18] According to Fernández Reyes (1993), the relationship between the environmental movement and the regional government has gone through three different stages. The first was characterized by the proximity between the two actors (the golden age), the second was one of continuous confrontation (that coincides with declining interest in social participation on the part of the administration), and the current one, which oscillates between agreement and confrontation.

In 1989, the Andalusian regional government approved the Protected Natural Sites Network by means of Law 2/1989. Integrated into the network was the Doñana Nature Park,[19] which covered 53,709 hectares and protected most of the surrounding area of the existing national park. Alongside Doñana, the Andalusian government also set up another 23 nature parks, 21 semi-urban parks (*parques periurbanos*), 32 nature spots (*parajes naturales*), 2 protected landscapes (*paisajes protegidos*), 35 natural monuments, 28 natural reserves and 4 concerted natural reserves (*reservas naturales concertadas*), not to mention the two national parks, Doñana and Sierra Nevada, which were also located in the region. This ambitious policy has meant not only that the region possesses the largest protected territory (over 19%) in Spain, but also that the Andalusian network of natural sites has become 'the most important one in the EU in terms of the number of areas protected and the surface covered by them' (Junta de Andalucía 2006) (Table 6.1).

According to the 1989 law, the regional Department of the Environment had to work out both a Plan for the Management of Natural Resources (*Plan de Ordenación de Recursos Naturales*, PORN) and a PRUG for Doñana.[20] However, these only became public in 1997 (*reglament* D 97/2005). A *reglament* from 1990 also stated that all nature parks had to be 'governed' by a ruling commission called the Junta Rectora. Finally, Law 2/1989 was modified by Law 18/2003, which introduced a new instrument of protection, the Areas of Community Importance (*Zonas de Importancia Comunitaria*, ZIC's). Since then, Doñana has become a ZIC because it belongs to the EU Natura 2000 Network, being both a special area of conservation and an area of special protection for birds.

In 1998, the worst ever ecological disaster affecting Doñana took place, the breaching of a nearby dam that contained toxic sludge.[21] The sludge flow bordered both the national and the nature park, entering the ecosystem in various ways. Since 1994, the environmental movement had repeatedly warned about the possibility of the dam collapsing, and one year later, a worker from the Canadian-owned mining

---

[19] The definition of a nature park is that of natural sites, little transformed by human activities or settlement, that, on the grounds of the beauty of their landscapes, the significance of their ecosystems or the uniqueness of their flora, fauna or geomorphological formations, possess ecological, esthetic, educational and scientific values whose preservation deserves preferential treatment. The only difference between nature parks and national parks is that the latter are declared to be of general interest for the nation.

[20] The PORN comes before the PRUG. The PORN embraces the following: an analysis of the state of conservation of resources and ecosystems in the area; the prohibitions to be imposed in view of the state of conservation; the specification of the adequate legal figure of protection for the area; the promotion of conservation, restoration and improvement measures; the elaboration of guidelines for sectoral policies and the planning of those economic and social activities that are compatible with the conservation of the site. The PRUG spells out the content of the PORN and is generally revised every eight years.

[21] For a detailed account of the disaster, see Díez Cansedo (2005). According to the author, if the EU directive on environmental liability (dir. 2004/35) had been in existence when the Doñana accident took place, the judge's decision, which acquitted the mining company, would have been different.

Table 6.1 Nature parks in Andalusia (Spain)

Nature parks	Province	Protected surface (hectares)
Sierra Nevada	Almeria, Granada	85,621
Cabo de Gata-Níjar	Almeria	45,663
Sierra María-Los Vélez	Almeria	22,670
Doñana	Cádiz, Huelva, Seville	53,709
Los Alcornocales	Cádiz, Málaga	167,767
Sierra de Grazalema	Cádiz, Málaga	51,695
Bahía de Cádiz	Cádiz	10,552
De la Breña y Marismas del Barbate	Cádiz	4,683
Del Estrecho	Cádiz	18,931
Sierra de Cardeña y Montoro	Córdoba	38,449
Sierra de Hornachuelos	Córdoba	60,032
Sierras Subbéticas	Córdoba	32,056
Sierras de Tejeda, Almijara y Alhama	Granada, Málaga	40,663
Sierra de Baza	Granada	53,649
Sierra de Castril	Granada	12,265
Sierra de Huétor	Granada	12,128
Sierra de Aracena y Picos de Aroche	Huelva	186,287
Despeñaperros	Jaén	7,649
Sierra Mágina	Jaén	19,961
Sierra de Andujar	Jaén	74,774
Sierras de Cazorla, Segura y Las Viñas	Jaén	209,920
Montes de Málaga	Málaga	4,996
Sierra de las Nieves	Málaga	20,163
Sierra Norte de Sevilla	Seville	177,484

company Boliden was sacked after revealing that this structure was in a state of disrepair. Following an angry dispute between the national and the regional authorities, which was clearly tainted with partisanship (the national government being in the hands of the Popular Party whereas the regional administration was socialist), as to which was responsible for the disaster, both of them produced different strategies in relation to the restoration of Doñana. As a result, the Doñana 2005 Project was put forward by the Spanish government and the Green Passageway (*Corredor Verde*) Project by the region.[22] Although the two projects shared a holistic and global approach, the fact that the two territorial levels responsible for the site, came up with different proposals was just another proof of a long-lasting tense relationship that had only contributed to aggravating the problems surrounding Doñana. It is true that the mixed management commission for the Doñana National Park, at which the national and regional authorities had to sit together, did not begin to work until 1999, but this should not excuse the absence of at least an informal relationship between

---

[22] Unless otherwise specified, developments in Doñana after the ecological disaster mainly draw upon Van der Zouwen (2006).

the two operating units in the area: the national-run Patronato for the national park, and the regional-run Junta Rectora for the nature park.

Although the working of the mixed-management commission, in theory, should have eliminated the anomaly expressed in the lack of fluent contacts between the two administrations, it has undoubtedly promoted a further reduction of social participation in the decision-making process concerning Doñana. 'In that respect the re-establishment of the national-regional relationship took place at the expense of the relationship between the governmental and non-governmental actors' (Van der Zouwen 2006:122). What is even worse, the new commission has not automatically generated inter-territorial cooperation, something that has been openly denounced by environmental associations. For instance, it was not until 2001 that a coordinating commission was set up to link the two different restoration projects advocated by the national and regional administrations. From that moment on, however, conservationists and authorities from both levels seem to have reached some common ground in terms of the discussion of appropriate restoration and management strategies for Doñana.

In 1992, the President of the regional government, Manuel Chaves, set up an international commission of experts with the aim of reaching a consensual and long-term solution for Doñana.[23] The main triggering factor behind this political move was the ecologically-threatening, but finally rejected, Costa Doñana project, which was announced at the end of the 1980's and which would have entailed the building of several facilities for around 20,000 people at a nearby location west of the protected site. In view of initial regional inaction in relation to the building scheme, concerted environmental pressure, both at the national and international levels, was organised to stop the project. Following a 1992 decision from the Andalusian High Court of Justice that denied the validity of the project, the regional government decided that it was about time to find an all-encompassing and durable solution for Doñana. The commission of experts that was called upon then produced the Sustainable Development Plan (SDP) for Doñana and its surroundings. The SDP, which counted on the participation of a wide array of actors (experts, public authorities, municipalities, environmentalists, productive interests), has been structured through different programmes, the most important of which are integral water management, farming practices, environmental protection, tourism, transport infrastructures, promotion of economic activities and cultural patrimony. Due to the coordination and implementation difficulties associated with the SDP, the regional government decided to authorize the Department of the Environment to set up the '21 Doñana Foundation', an agency in charge of putting the SDP into practice (Fundación Doñana 21 2006). Furthermore, this foundation was established to offset criticisms about the lack of a global and comprehensive approach in relation to the

---

[23] Leaving aside the mining spill, Doñana has constantly been under threat. Cattle-breeders have on certain occasions invaded Doñana and let their cattle graze on the protected area. Countless bird and fish death cases have also been detected, related to the use of pesticides by nearby farmers. More recently, the water works in the Guadalquivir river have impinged on the quality and quantity of the water reserves in Doñana, to cite just a few examples.

application of the different sectoral programmes that the SD strategy contained. The
need for a more participatory forum at which to discuss the future of Doñana was
yet another reason behind the establishment of the foundation.

At present, Doñana continues to be in the spotlight. The dangers surrounding
its fragile ecosystem frequently make the headlines. The most recent problems affecting the site have been the increasing amount of illegal hunting and a worrying
growth in drug-smuggling, the latter taking advantage of the nearly 50 km of wild
beaches bordering the site to the south (Los Verdes de Andalucía 2006). Moreover,
tension between conservation needs and production activities in the vicinity seem
to have become endemic to Doñana management, regardless of the participatory
mechanisms or intergovernmental arrangements that are being put into practice. As
a result, the difficulties associated with the idea of SD come most visibly to the fore
when discussing the present and future of Doñana.

6.6 Conclusions: Legitimacy and the Sustainable Development Dilemma

Advocates of participatory democracy may concur with the idea that all interested
parties in nature policy should be encouraged to partake in its decision-making process. Critical authors with normative definitions may contend that democracy is basically a mechanism for choosing governments rather than different nature policies,
with the choice about which courses of action to adopt then being the responsibility of the governments we have chosen (Sartori 2003). Although the latter line of
thought takes into account that 'good' democracy must embrace both a normative
(what it should aspire to) and a realistic (how it works in practice) dimension, it
might nonetheless consider that the second dimension (or the emphasis on constitutional checks and balances, and the rule of law) is more relevant than the first
one (that stresses the virtues of participation and responsiveness). If nature policies
adopted by democratically elected governments are accompanied by social involvement, transparency and accountability, then the realistic and normative dimensions
of democracy might ideally converge. Yet, if the participatory element were missing,
this would not immediately lead to the disqualification of the policy adopted by the
government.

The situation is, however, not as simple as that because Article 6 of the 1997
Amsterdam Treaty mandates that environmental protection requirements must be
integrated into the definition and implementation of European Community policies,
in particular with a view to promoting SD. That is to say, even democratically
elected governments do not have a free hand when opting for any specific type
of nature policy. Moreover, these governments have also committed themselves in
the international arena to applying SD strategies that envisage two elements, certain types of nature policies (policies that efficiently tackle specific collective problems, as expressed by output-oriented legitimacy) and certain types of guidelines
about how to elaborate and implement those policies (policies that allow for social

participation while endorsing transparency and accountability as procedural mechanisms, as expressed by input-oriented and throughput-oriented legitimacy).

In the end, the question to be addressed is: Are democratic regimes adequately prepared to tackle the new SD challenge? As Lundqvist (2004) rightly points out, 'There is an inherent conflict between political time frames and ecological lifecycles. Short-term tactics and strategic timing may lead to a smoother functioning of the political processes of the day, but detract from the longer-term viability and productivity of our ecological life-support systems' (2004: 96). Should nature policy ever become sustainable, it will need a determined and fully committed political mandate that transcends mutable partisan politics and electoral cycles. Equally true, SD cannot be left to endless, time-consuming, resource-absorbing political games that are based on lowest-common-denominator interest groups. However, SD cannot be pursued in the absence of democracy either, as all international programs have clearly established.

All these dilemmas have been troubling Doñana since it was originally preserved in the 1960's. When Doñana was first established, lack of input legitimacy could partly be justified by reasons of 'ecological urgency' and explained by the peculiar political circumstances under which Doñana became a public issue. Furthermore, regardless of the political context in question, actions concerning Doñana were undoubtedly needed for the sake of its preservation. Later on, under democratic circumstances, Doñana was still in peril and demanded, therefore, new strategies for its protection. That these strategies were, in most cases, adopted without a rigorous participatory process can obviously be criticized. However, previous experiences of social involvement in regional nature policy had also involuntarily led to certain production interests gaining prominence to the detriment of conservation and an explosion of social conflict.[24] The lessons to be drawn from these conflictive experiences are that broad social participation (input legitimacy) and even impeccable environmental governance (throughput-oriented legitimacy) do not necessarily produce good policies (output legitimacy), or policies that are capable of satisfying current needs without impinging upon the needs of future generations. The fact that a general consensus exists about the need for the SD challenge to be linked to participation makes things even more complicated because the possibility of democratic governments then taking good sustainable decisions without consulting the interested parties – which might be called 'enlightened despotism' – cannot be advocated.

---

[24] In the real world, the average citizen does not seem to be very keen on participating in public affairs. As a matter of fact, those groups that strongly care about taking part might represent powerful and well-articulated pro-development interests that clearly collide with conservation strategies, in particular, or with the SD agenda, in general.

References

Aguilar Fernández, S. (1997a). 'Abandoning a Laggard Role? New Strategies in Spanish Environmental Policy'. In M. Skou and D. Liefferink (eds) *The Innovation of EU Environmental Policy*. Copenhagen: Akademisk Forlag/Scandinavian University Press.

Aguilar Fernández, S. (1997b). *El Reto del Medio Ambiente. Conflictos e Intereses en la Política Medioambiental Europea*. Madrid: Alianza Universidad.

Aguilar Fernández, S. and M. Jiménez (1999). 'Las Marismas de Odiel'. In S. Aguilar, N. Font and J. Subirats (eds) *Política Ambiental en España*. Valencia: Tirant Lo Blanch.

Aguilar Fernández, S. (2003). 'Spanish Coordination in the EU: The Case of the Habitats Directive.' *Administration & Society*, 34(6): 678–699.

Aguilar Fernández, S. (2004). 'Spain: Old Habits Die Hard'. In A. Jordan and D. Liefferink (eds) *Environmental Policy in Europe. The Europeanization of National Environmental Policy*. London: Routledge.

Díez Cansedo, G.S. (2005). 'La Responsabilidad Ambiental y los Riesgos Ambientales en el Desarrollo de Actividades Productivas. Reflexiones a Partir del Caso del Derrame de Lodos Tóxicos en las Minas de Aznalcóllar'. Final report for the Master on Environment (Carlos III University, Madrid).

Elías Mortera, L.M. (1999) 'Las Políticas Forestales en México y España'. Doctoral thesis, Ortega and Gasset Institute (Madrid). Supervisor: Susana Aguilar Fernández.

Fernández Reyes, R. (1993). 'El Ecologismo en el Ecosistema Social Andaluz.' Consejería de Medio Ambiente, Publicaciones, Seville, Revista *Medio Ambiente*.

Figueroa Acosta, C. and Toro García, P. (2003). 'Desastre en Doñana'. Final Report for the Workshop on Environmental Conflicts. Master on Environment (Carlos III University, Madrid).

Fundación Doñana 21 (2006). 'Desarrollo sostenible' (http://www.donana.es/nueva/donana.php).

Junta de Andalucía (2006). 'La Red de Espacios Naturales Protegidos.' (http://www.juntadeandalucia.es/medioambiente/espacios_naturales)

Jordan, A. and D. Liefferink (eds) (2004). *Environmental Policy in Europe*. London: Routledge.

Los Verdes de Andalucía (2006). 'Educación. Noticias.' (http://www.losverdesdeandalucia.org/noticia.php).

Lundqvist, L.J. (2004). 'Management by Objectives and Results: A Comparison of Dutch, Swedish and EU Strategies for Realizing Sustainable Development'. In W.M. Lafferty (ed) *Governance for Sustainable Development*. Cheltenham, Gloucestershire: Edward Elgar.

Meadowcroft, J. (2004). 'Participation and Sustainable Development: Modes of Citizen, Community and Organisational Involvement.' In W.M. Lafferty (ed) *Governance for Sustainable Development*. Cheltenham, Gloucestershire: Edward Elgar.

Quercus (environmental magazine), Madrid, various issues, 1998–2003.

Romero, J. (2005). 'Capitalismo de casino'. *El País* (22 December 2005).

Sartori, G. (2003). *¿Qué es la Democracia?*. Madrid: Taurus.

Scharpf, F.W. (1998). *Interdependence and Democratic Legitimation*. Cologne: Max Planck Institute for the Study of Societies.

Zouwen, M. Van der (2006). *Nature Policy Between Trends and Traditions*. Delft: Eburon.

Chapter 7
How to Deal with Legitimacy in Nature Conservation Policy?

Jan van Tatenhove

7.1 Introduction

In her chapter 'Legitimacy Problems in Spanish Nature Policy: The case of Doñana' Susana Aguilar Fernández addresses different forms of legitimacy and the dilemma between legitimacy and sustainable development in nature conservation policy. The chapter discusses the institutionalization of nature conservation policy in Spain, especially nature conservation policy in Doñana. Doñana, one of the jewels in the crown of European biodiversity, was designated as a national park at the end of the 1960's and as a nature park at the end of the 1980's. During this period both national and regional governments were responsible for conservation in the Doñana area. After the accession of Spain to the EEC (later the EU) in 1986, conservation policy in Doñana was more and more influenced by European and international rules and regulations, such as the Ramsar Treaty and the EU Birds and Habitat Directive. An important renewal of policy took place after the ecological disaster of 1998, when a dam broke and toxic sludge contaminated the fragile biodiversity of Doñana (for an extensive analysis of nature policy in Doñana see, Van der Zouwen 2006).

Although the theme of Aguilar Fernández's chapter is legitimacy problems in Spanish nature policy she does not make clear what kind of legitimacy problems nature conservation policy faces and what sort of relationship exists between input, output and throughput legitimacy. In her concluding section Aguilar Fernández discusses the tension between good democracy and the requirements of sustainable development. According to Aguilar Fernández democratically-elected governments do not have a free hand when opting for any specific type of nature policy, because they have committed themselves in the international arena to apply sustainable development strategies that envisage two elements: 'certain types of nature policies (policies that efficiently tackle specific collective problems, as expressed by output-oriented legitimacy) and certain types of guidelines about how to elaborate and implement those

Jan van Tatenhove
Environmental Policy Group, Department of Social Sciences, Wageningen University and Research Centre
Email: jan.vantatenhove@wur.nl

J. Keulartz and G. Leistra (eds.), *Legitimacy in European Nature Conservation Policy: Case Studies in Multilevel Governance*, 101–108. © Springer 2008

policies (policies that allow for social participation while endorsing transparency and accountability as procedural mechanisms, as expressed by input-oriented and throughput-oriented legitimacy)' (Aguilar Fernández this volume, p. 99).

She asks herself the question: 'are democratic regimes well prepared to tackle the novel sustainable development challenge?' In other words how do governments deal with the tensions between the involvement and participation of actors in the development and implementation of nature conservation policy, while at the same time realizing a sustainable form of nature conservation, which might conflict with the interests of the actors involved. By answering her question she states the following considerations. Firstly, if nature policy is ever to become sustainable, it needs a determined and fully committed political mandate that transcends changeable partisan politics and electoral cycles. Secondly, sustainable development cannot be left to endless, time-consuming, and resource absorbing and lowest-common denominator interest group based political games. Thirdly, sustainable development cannot be pursued in the absence of democracy either, as all international programs have clearly established. From these dilemmas – illustrated with the Doñana case – Aguilar Fernández draws the following lessons: broad social participation (input legitimacy) and even impeccable environmental governance (throughput legitimacy) do not necessarily produce good policies (output legitimacy), or policies that are capable of satisfying current needs without impinging harmfully upon the needs of future generations. 'The fact that a general consensus exists about the need for the sustainable development challenge to be linked to participation makes things even more complicated because the possibility of democratic governments then taking good sustainable decisions without consulting the interested parties – which might be called "enlightened despotism"- cannot be advocated.' (Aguilar Fernández this volume, p. 99).

Aguilar Fernández formulates the sustainable development challenge as a policy paradox (compare Stone 2002). Sustainable development can best be achieved by enlightened despotism, while at the same time sustainable development needs new forms of environmental governance to enable the (in)direct participation of the stakeholders involved. However, Aguilar Fernández offers no solution for this policy paradox, which she frames as the tension between output legitimacy on the one hand and input and throughput legitimacy on the other. What can be said about Aguilar Fernandez' assumptions? Does participation really hamper a sustainable nature policy? Are input and throughput legitimacy on the one hand and output legitimacy on the other incompatible in realizing sustainable nature policy? How can one theoretically understand the relationships between the different forms of legitimacy in relation to the institutionalization of a sustainable nature conservation policy in the context of European governance?

In this comment I will elaborate upon dimensions of legitimacy in the context of the institutionalization of nature conservation policy and sustainable development. After discussing the concept of legitimacy, I will turn to the specific characteristics of nature conservation politics in the context of European governance. This provides me with the tools to understand the specific characteristics of legitimacy in nature conservation policy. By introducing the framework of the 'staging of practices'

(Van Tatenhove et al. 2006) it is possible to formulate some general assumptions about legitimacy and sustainable nature conservation policy in a multi-level context.

7.2 Legitimacy

Legitimacy is a central concept in the study of politics and policy. In general, legitimacy refers to the acceptance of law, law-making or rules, but is much broader than mere legality. While legality denotes whether a rule was made correctly, that is, following regular procedures, legitimacy refers to whether people accept the validity either of a specific law or of the entire political system (Hague and Harrop 2001). Scharpf (1999) elaborated on legitimacy in the specific policy context of the European Union. He made a distinction between input-oriented legitimacy and output-oriented legitimacy.

Input-oriented legitimacy emphasizes 'government by the people'. Political choices are legitimate if they reflect the 'will of the people' – that is, if they can be derived from the authentic preferences of the members of a community. Input-oriented legitimacy thus refers to the reflection of the interests of participants involved in the formulation of politics and policy. Participation can vary from constitutionally institutionalized and supplementary forms of participation (elections, public hearings, citizens forums) to more reflexive or deliberative forms of participation (interactive policy making, authentic dialogue) (Van Tatenhove and Leroy 2003; Innes and Booher 2003; Hajer and Wagenaar 2003).

Output-oriented legitimacy emphasizes 'government for the people': political choices are legitimate if and because they effectively promote the common welfare of the constituency in question (Scharpf 1999: 6). '"Government for the people" derives legitimacy from its capacity to solve problems requiring collective solutions because they could not be solved through individual action, through market exchanges, or through voluntary cooperation in civil society' (Scharpf 1999: 11). According to Scharpf in democratic nation-states, input- and output-oriented legitimacy coexist side-by-side, reinforcing, complementing and supplementing each other. The EU on the other hand only aspires to output-oriented democratic legitimacy, i.e. interest based rather than identity based. The consequence of this is that there is no reason to assume solidarity among the members of the constituency and there is less reason to assume that the actors in charge of making collectively binding decisions will single-mindedly and effectively pursue the public interest.

In political science the traditional dilemma between input and output legitimacy is how to increase, for example, the effectiveness of policy without limiting participation and popular consent, by increasing the transparency and accountability of procedures (throughput legitimacy). However, as we have seen, input, throughput and output legitimacy are different processes, with different dynamics in national political systems and European politics. But legitimacy depends not only on the political system or the public policy process itself, but also on the specific domain of politics, in this case nature conservation policy. Therefore in the next section I take a look at some of the characteristics of nature conservation.

7.3 Characteristics of Nature Conservation Policy

Nature conservation policy has developed from the protection and conservation of nature areas by private parties and governments at the end of the 19th and the beginning of the 20th century to nature development and the creation of ecological networks through Europe in the 1990's. With the introduction of the concept of Natura 2000, the European Union broadened its policy scope with regard to nature conservation. Natura 2000 envisages the establishment of a network of conservation areas throughout the territory of the member states. Crucial to this is the protection of specific habitats, i.e. ecological communities of flora and fauna bound together by particular physical circumstance. Besides the importance of habitats, the Natura 2000 concept also emphasizes the assessment of sites on the basis of biogeographical regions, the need for core area and buffer zone protection and the problem of connecting these two to achieve an ecological network, while maintaining specific human activities within special areas.

Nowadays, nature conservation is a highly dynamic field of policy. The dynamic has both a governance and a multi-level governance character (Van der Zouwen 2006). On the one hand there is the emergence of more levels (EU, national states, and regions) and shifting interrelations between these levels in nature policy, on the other hand new actors are getting involved in nature policy processes, especially at the local level. According to Van der Zouwen (2006: 25–26) governance and multi-level governance manifest themselves in an increasing interconnectedness between EU, national, regional and local levels of policy making, in the emergence of network-like structures, in which public and private actors from different territorial levels interact, and in shifts in the locus of policy making towards informal practices as well as sub-national and transnational practices and towards ad-hoc and temporary coalitions.

Another important characteristic of nature policy is its discursive and scientific basis. Not only is nature conservation policy strongly influenced by ecological and biological concepts and theories, such as the island theory and system ecology, but the institutionalization of nature conservation policy is also strongly influenced by nature discourses. Each member state developed specific nature arrangements, based on specific discourses on nature, coalitions, resources and rules of the game (Liefferink et al. 2002; Van der Zouwen 2006). For Doñana, Van der Zouwen describes three discourses 'territorial water management', 'water system and ecosystem' and 'sustainable development' and she analyses how the advocacy of new policy discourses results in the emergence of new nature policy arrangements. Scientific knowledge also plays a crucial role in nature policy. A wide range of biological, ecological, entomological, environmental and spatial planning knowledge is framed and translated into policies (Liefferink et al. 2002). In this process of framing, institutes and organizations, such as the European Environmental Agency, biogeographical seminars and the Environment Information and Observation Network (EIONET),[1] fulfil important roles, in different phases of the policy-making process, such as agenda-setting, policy formulation and implementation.

Both Aguilar Fernández (this volume, pp. 83–100) and Van der Zouwen (2006) analyse how the highly dynamic governance and multi-level governance contexts have influenced the formulation and implementation of nature conservation policy in Doñana. Aguilar Fernández illustrates the tension and fragmentation of nature conservation policy in Spain, especially the struggle of competencies between ministries, the relationship between national government and regional government and local actors and the interplay of the national and regional level with the EU-level. This case study is an illustration of nature conservation policy as a dynamic field of policy, in which agenda setting, policy formation and implementation are shaped by multi-actor and multi-level setting. More specific national and local initiatives have to deal with the openness and accountability of the different stages of the decision-making process in the EU; the empowerment of the various actors in the system and the basis of such power; and finally the problem solving capacity of the EU political system. The continuously shifting balance between domestication and Europeanization becomes an ever more pressing issue, given the complicated relationships of coordination and authority between the different levels and the supranational and non-governmental actors in EU nature conservation policy. This is particularly so in view of the serious legitimacy crisis presently facing the EU and requiring even better coordination, both horizontally across policy sectors and vertically between the EU and its member states.

7.4 The Consequences of the 'Staging of Practices' for Issues of Legitimacy

As stated before, Aguilar Fernàndez discusses the relationship between legitimacy and nature conservation only in a superficial way. Her main conclusion is that participation and environmental governance do not necessarily produce good policies. Is there more to say than she does? Under what conditions will participation produce legitimate and sustainable nature conservation policy? Is participation really a complicating factor for sustainability or is it possible to bring input and output legitimacy together? The central question of this section will be how we can understand the relationship between input, throughput and output legitimacy given the multi-level governance character of EU nature conservation policy. To answer this question I will start with the example of the U.S. Endangered Species Act in which input, throughput and output legitimacy are brought together.

The Endangered Species Act[2] – comparable with the EU Habitats Directive – prohibits the 'taking – killing or injuring – of any wildlife listed as an endan-

---

[1] EIONET is a collaborative network of the European Environmental Agency and its Member states, connecting National Focal Points in the EU, European Topic Centres, National Reference Centres and Main Component Elements (see Buunk 2003).

[2] The case description is based on Fung and Wright (2001).

gered species through either direct means or indirect action such as modification of its habitat'. Not only did this law only protect listed species, it stopped
productive development projects. The introduction of the Habitat Conservation
Plan (HCP) created the option to escape deadlocks. In a deliberative process,
officials of local and national environmental agencies, developers, environmental activists, and community organization, developed HCP's in which environmental and
economic objectives were met. This case shows that it is possible to overcome deadlock situations by introducing innovative methods of deliberation and participation
to realize an effective and legitimate sustainable nature conservation policy.

Given the specific characteristics of nature conservation policy in Doñana, it is
also necessary to find innovative solutions for nature conservation policy in the
European context. These characteristics include the interplay of processes of decentralization and Europeanization, problems of coordination, unclear institutional
frameworks, and competency struggles between departments, and differing roles
for NGO's (at several levels). More generally, to understand problems of legitimacy
in nature conservation policy in Doñana, one has to understand the characteristics
of European governance, the specific transnational arrangements which influence
nature conservation policy in Doñana, and the interplay of formal and informal
practices within these arrangements.

The EU is in a state of constant flux, fragmented and balancing between intergovernmental bargaining and multi-level governance. This makes policy-making
and politics in the EU unique compared to, for example, nation-states. To understand
nature conservation policy making in such a multi-level context and to formulate
some preliminary assumptions about forms of legitimacy in sustainable nature conservation policy, the framework of the 'staging of practices' (Van Tatenhove et al.
2006) could be useful.

European governance can be analyzed as the 'staging of practices' as a way to
understand the interplay of formal and informal practices. Informal practices are
'those non-codified settings of day-to-day interaction concerning policy issues, in
which the participation of actors, the formulation of coalitions, the processes of
agenda-setting, (preliminary) decision-making and implementation are not structured by pre-given rules or formal institutions' (Van Tatenhove et al. 2006: 17). Informality can take place in 'front-stage' or 'backstage' settings. 'Front-stage' refers
to policy making in rule-directed arrangements. Central to these arrangements is
policy making as the interplay primarily between EU and member-state institutions.
The Treaties provide the basic set of rules for decision-making procedures and the
competencies of the actors involved. Backstage politics is the setting in which the
roles of actors and the rules of the game are not established beforehand, nor structured by the rules and norms of existing institutions. Backstage is concerned with
(rule-altering) arrangements evolving 'on the ground', in which participants have
the opportunity to change the rules of the game (Van Tatenhove et al. 2006: 16). At
the same time these settings can be formally sanctioned or not (see Table 7.1).

Besides formal frontstage practices, three types of informal practices can be distinguished: informal frontstage (based on rules that are not formally codified and
sanctioned), formally sanctioned backstage (practices initiated and authorised by

Table 7.1 The staging of practices: formal and informal settings in EU politics (Van Tatenhove et al. 2006: 17)

Settings	Formally sanctioned	Not formally sanctioned
Front stage (rule-directed; pre-give rules)	1. Formal front stage	2. Informal front stage
Back stage (rule-altering; no pre-given rules)	3. Formally sanctioned backstage	4. 'Sub-politics'

formal institutions, but at the same time have a 'rule-altering' nature) and sub-politics (practices with no pre-given rules and without formal basis).

Given the characteristics of policy making in a multi-level setting, only detailed analyses of the staging of practices make it possible to determine the relationship between input, throughput and output legitimacy in specific cases of nature policy and to develop 'the tools' to find answers for the paradox between input and output legitimacy in sustainable nature policy projects. Attention should be paid to questions such as who decides about issues of inclusion and exclusion (who is allowed to participate?), who has the resources (for example scientific knowledge) to define the rules of the game? How do different discourses influence participation and outcomes? More generally, this kind of analysis can deepen our understanding of the openness of policy processes (both in terms of access and information), accountability and transparency of the different stages of the decision-making process in the EU.

References

Buunk, W. (2003). *Discovering the Locus of European Integration. The Contribution of Planning to European Governance in the Cases of Structural Fund Programmes, Trans-European Networks, Natura 2000 and Agri-Environmental Measures.* Delft: Eburon.

Fung, A. and E.O. Wright (2001). 'Deepening democracy: Innovations in empowered participatory governance.' *Politics & Society* 29(1): 5–41.

Hague, R. and M. Harrop (2001). *Comparative Government and Politics. An Introduction.* New York: Palgrave.

Hajer, M. and H. Wagenaar, eds. (2003). *Deliberative Policy Analysis. Understanding Governance in the Network Society.* Cambridge: Cambridge University Press.

Liefferink, J.D., J.P.M. van Tatenhove and M.A. Hajer (2002). 'The dynamics of European nature policy: The interplay of frontstage and backstage by 2030'. In W. Kuindersma, ed. *Bestuurlijke trends in het natuurbeleid. Planbureaustudies, nr.3.* Wageningen: Natuurplanbureau, pp. 39–65.

Innes, J.E. and D.E. Booher (2003). 'Collaborative policymaking: Governance through dialogue.' In M. Hajer and H. Wagenaar, eds. *Deliberative Policy Analysis. Understanding Governance in the Network Society.* Cambridge: Cambridge University Press, pp. 33–59.

Scharpf, F. (1999). *Governing in Europe. Effective and Democratic?* Oxford: Oxford University Press.

Stone, D. (2002) (Revised Edition). *Policy Paradox. The Art of Political Decision Making.* New York/London: W.W. Norton & Company.

Tatenhove, J. van and P. Leroy (2003) 'Environment and participation in a context of political modernisation.' *Environmental Values* 12: 155–174.

Tatenhove, J. van, J. Mak and D. Liefferink (2006). 'The interplay between formal and informal practices.' *Perspective on European Politics and Society* 7(1): 8–24.

Zouwen, M. van der (2006). *Nature Policy between Trends and Traditions. Dynamics in Nature Policy Arrangements in the Yorkshire Dale, Doñana and the Veluwe.* Delft: Eburon.

Bergo-Malax Outer Archipelago (by Carina Jarvinen)

Chapter 8
Resistance to Top-Down Conservation Policy and the Search for New Participatory Models

The Case of Bergö-Malax' Outer Archipelago in Finland

Susanna Björkell

8.1 Introduction

As a response to the fact that implementing Natura 2000 in Finland has proven problematic, we would like to elaborate on the possibilities of including local actors in the management of nature resources. In so doing, the social and economic aspects of sustainable development that were lost during the interpretation of the Natura 2000 directives in Finland could be re-introduced. We are suggesting the theories of common-pool resource management and co-management in order to highlight an alternative to the separation of nature and humans. The common-pool theory claims that the local user's lack of influence on the methods for resource management increases the risk of failure for co-management. According to our observations, in the case study in Bergö-Malax' outer archipelago, we discovered important local management systems, which will play a vital role in building up the potential for success for the management of the area. Our analysis emphasises the mandate for civil servants to make sovereign decisions with local level actors during the planning process. If this mandate did not exist, it is argued that meetings between the authorities and the local actors would be more informative events than ones in which the actors have true influence, and would consequently have less chance of enhancing the legitimacy of the process.

Susanna Björkell

Centre for Administration and Organisational, Regional and Environmental studies, FO-RUM
Helsinki University
E-mail: susanna.bjorkell@helsinki.fi

J. Keulartz and G. Leistra (eds.), *Legitimacy in European Nature Conservation Policy:*
Case Studies in Multilevel Governance, 109–126. © Springer 2008

8.2 Implementing Natura 2000 in Finland

Natura 2000 is the EU's nature protection network. It differs from earlier Finnish nature protection programmes because it is stricter and more widespread. Natura 2000 became a contentious issue when Finland entered the EU. Some even called it the largest environmental conflict in Finland's history. The main reasons for its being so controversial can be said to have been caused by the lack of legitimacy in its implementation process. During the first years of the implementation of Natura 2000, conflicts occurred between environmental authorities and local landowners, because decisions about which areas to include in the Natura 2000 network were quickly made and did not always convince the local actors. Due to lack of time and resources, communication between the authorities and the locals was infrequent. Only in some cases were landowners individually informed about their land becoming part of the Natura 2000 network. All together, landowners sent in 15,000 legal complaints to the Finnish government opposing Natura 2000. This action by the landowners clearly indicates a lack of legitimacy on the part of the Natura 2000 network. In order to settle the conflict between the environmental authorities and the landowners, decision-making was transferred from the environmental authorities to the national government, further illustrating the severity of the conflict (see Malmsten 2004; Nieminen 2000; Bonney 2004; Korhonen 2004 for more information on the Finnish implementation of Natura 2000).

In their attempt to implement Natura 2000, the environmental authorities acted in a way that can be described as strictly top–down. This was not suitable for the situation. In the meantime, the Finnish government has recognized the external factors that increased the potential for conflict. In short, the agrarian and rural community in Finland has an aversion to the EU, which stems from the weakened prospects of Finnish farmers in the European agricultural market. The aversion can also be explained by the strong landowning tradition, which exists in Finland. There is a common perception about the sovereignty of the landowner which is difficult to reconcile with the environmental regulations of the EU, because the EU requires that nature protection should apply to state owned as well as privately owned land. The conflict that emerged during the implementation of Natura 2000 can partly be understood within this context of public versus private land ownership. Needless to say, the public meetings that were arranged to discuss Natura 2000 in Finland became opportunities to vent frustrations not only about Natura 2000 but also about the EU in general (Malmsten 2004). In Finland, the agrarian and environmental authorities did not form a consensus about the Natura programme. The interest organizations (the farmers' union and the Finnish association for nature conservation), however, managed to mobilize public opinion and dominated the media, which increased concerns of the landowners affected (Malmsten 2004).[1]

With the lessons learned from the conflicts of the initial phase of Natura 2000, there is currently a drive among environmental authorities to find new ways to

---

[1] Jokinen (1995) describes the development of agro-environmental policy in Finland).

implement nature conservation policies in a more legitimate way. The discussion about increasing stakeholder participation has started at most levels of the environmental administration. Efforts have been made to re-gain the trust that was lost in the implementation of Natura 2000. Openness and public participation are today seen as a precondition for nature protection. Because of the growing emphasis on protecting areas affected by humans such as traditional agricultural regions, there is a great need for co-operation between the local communities and the environmental authorities when it comes to the practical management of resources. In order to do this, it is important to create a legitimate basis for encouraging co-operation. In this regard, co-management is a management model that can be used to build trust and to increase legitimacy.

8.3 Theoretical Framework of Co-management of Nature Resources

8.3.1 Co-management

Before the introduction of today's environmental administration, the community managed natural resources. The role of the government as the manager of natural resources has grown during the last century.[2] Because the impact of the measures for managing nature resources has become more severe for resource users and local inhabitants, the need to include them in the management of these resources has grown. Jentoft et al. (1998) describe the crisis in fisheries management. They think that crises in the management of nature resources have harmed the top–down approach and science-based decision-making. They claim that all models for fisheries management imply that there is a need for resource users to become more involved in the management process. One can say that top–down policy is not compatible with current social trends. Jentoft et al. (1998) state that the role of the government in fisheries management is characterized by distance, impersonality and an insensitive bureaucratic approach. Although the image of the government might differ between management systems and areas, it can be assumed that the characteristics mentioned above also illustrate to some extent, the image of the Finnish government not only in regard to fisheries management but also in regard to other nature resources. Including the locals by sharing power and responsibility in resource management would presumably improve that negative image. This participatory method is called co-management (Carlsson and Berkes 2003).

As a means for increasing legitimacy, co-management can be seen as an answer to the reactions and objections to the top–down steering model that has dominated environmental administration. The old nature protection administration was based on the assumption of the Tragedy of the Commons. In short, if users are given

---

[2] In 1923, the first law for nature conservation was established in Finland; in 1983, the Ministry of the Environment was founded.

open access to a resource, they will eventually overexploit it, which will result in the destruction of that resource. Managing nature resources with this assumption has been a self-fulfilling prophecy, since, when a community only regards itself as an aggregation of individuals, instead of as a functioning unity, the individuals of that community are only motivated to act opportunistically. Co-management regards resource users more as communicative community members than as non-communicative opportunists.

Borrini-Feyerabend et al. (2000) define co-management as a 'situation in which two or more social actors negotiate, define and guarantee amongst themselves a fair sharing of management functions, entitlements and responsibilities for a given territory, area or set of natural resources'. This definition implies that there is negotiation on an equal basis about how the power is shared among the actors. However, to some extent, it excludes situations where one actor historically dominates the management and offers to give away part of its monopoly of power (compared to the situation where the state dominates the management but wants to give away some of the power to the users). The term social actor can mean individual actors or groups of several individuals. In the definition, it is also understood that the term co-management is used to address issues concerning the management of natural resources.

The International Union for Conservation of Nature and Natural Resources (IUCN) defines co-management as 'a partnership in which government agencies, local communities and resource users, non-governmental organizations and other stakeholders negotiate, as appropriate to each context, the authority and responsibility for the management of a specific area or set of resources' (IUCN 1997). This definition shows the importance of taking into account not only the individual users but also the community they are part of. It also emphasises that the best form of co-management varies from case to case and that the role of NGO's is important for the community.

All definitions of co-management include both the private sector and the state. The relationship between the state and the private sector can vary from case to case and from one co-management system to another. Carlsson and Berkes (2003) present a model with four alternatives describing the interaction between public and private actors (See Fig. 8.1).

Figure 8.1a) shows a type of interaction that is characterized by fraternization between two separate spheres: the public and the private actors. This type of co-management includes the exchange of information as well as the exchange of goods and services. Figure 8.1b) shows two overlapping sectors. This figure suggests that co-management is a matter of intersecting intercepting parts of the spheres. This type of co-management emerges when state representatives and groups of resource users form joint management bodies or participate in joint decision-making. In this type of co-management, each sector keeps its relative autonomy and co-management can be regarded as a formal arena for cooperation. There are also non-governmental organisations formed between representatives of the state and the private sector. Here, the borders between the sectors are blurred, and it may not be possible to talk about separate spheres of authority. Figure 8.1c) shows a situation

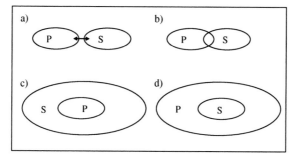

Fig. 8.1 Private sector and state interaction in co-management (Carlsson and Berkes 2003)

where the private sphere is nested inside the public. Here, the state has legal rights over the area or the resource, but the private actors can be entrusted with some rights to manage or use the state owned area or resource. There are cases where the private actors form a management organization with a substantial degree of independence. Figure 8.1d) illustrates the reverse situation where the state operates inside a non-public sphere, where the legal rights are in the hands of the private actors. The state can, however, put restrictions on the management of the resource. This type of co-management does not always constitute a source of objection by the private actors. There are cases where appropriators have a need for third-party solutions, which are often provided by the state.

This categorization by Carlsson and Berkes illustrates well the variety of co-management systems. In spite of the categorizations, it should be noted that the concept of co-management embraces a continuum from simple exchange of information to formal partnerships (Carlsson and Berkes 2003). Co-management also changes with time and evolves; thus, it should be seen as a process rather than as a static situation.

On the one hand, co-management can be regarded as something new, a response to the resistance to the traditional top–down management model. On the other hand, co-management can be traced back to traditional societies, where natural resources were managed through reciprocity and solidarity. According to Borrini-Feyerabend et al. (2000), 'these systems were embedded in the local cultures and accommodated for differences in power and roles – including decision-making – within holistic systems of reality and meaning'. Discussions between interested actors occurred normally and the dialogues were based on field experience. Thus, the structures of co-management can be found in historical nature resource management. Jentoft et al. (1998) claim that 'co-management is not so much a question of reinvention as of rediscovery and renewed commitment to the "meso-level" of governance, involving civil society and voluntary associations'. Borrini-Feyerabend et al. also state that many contemporary nature resource management situations show a mixture of old and new. Considering the importance of the historical structures of the community for co-management, it is important to acknowledge possible existing structures of self-regulation in the process of establishing co-management. 'Cultural and social qualities of human communities are assets that resource managers can draw upon and possibly reinforce', Jentoft et al. (1998) claim. They mean that co-management can, therefore, be a product of a vital community with social integration. However, they also emphasize the possibility that co-management can contribute to community vitalization and social integration.

It has been concluded that a sustainable community is a precondition for sustainable nature resources. Jentoft studies coastal fisheries and points out that it is not only the community that depends on the fish stocks but also the other way around; for sustainable fish stocks, there has to be a functioning community (Jentoft 2000). When community cohesion becomes weak, the individuals act more opportunistically, which leads to overuse of resources. The overuse of a resource can be a sign of community failure. The users see the resource as an externality instead of a common good, and the situation defined as the tragedy of the commons occurs (Hardin 1968). This situation shows the importance of taking community structures into account when planning the management of a nature resource. It is of great importance that governmental managers implement nature protection programmes in a sensitive way, applying the existing organisational features of the community instead of replacing them with top–down management regimes. Jentoft suggests that managers should adopt designs and methods that could rehabilitate and strengthen not only the community's solidarity but also its cultural qualities (Jentoft 2000).

There are two major lines of argument for co-management. Firstly, the outcome of the management process will be more valid because of the increased knowledge used during the process. The local users have experience-based knowledge complementary to the knowledge of civil servants who are not as familiar with local characteristics. Secondly, the legitimacy of management is increased with co-management processes. Through participation, openness and trust will grow and, as a result, the legitimacy of the management increases. In co-management, when users receive more responsibility over the management in functional terms, they are likely to act more responsibly in moral terms, which implies wide-scale compliance with the rules and regulations agreed upon (Jentoft et al. 1998).

There are also arguments against co-management of nature resources, some of which claim that co-management might imply a compromise to conservation goals. Local users having an impact on the management of a resource are suspected of not being able to value and preserve the resource without ulterior motives. However, as Jentoft et al. (1998) have concluded, individuals do not only act out of self-interest. The roles they have as members of the community, social groups and organizations also influence their interests.

New directions for nature resource management require that officials and planners acquire some new social skills for facilitating negotiations between the various actors. For example, it is important that the civil servant who represents the environmental authorities has the ability to listen well to all opinions, while remaining objective. He or she also has to be able to show respect for the participants and have a general ability to understand people (Turunen 2005). The basic condition for successful co-management is that the civil servant working with the management in co-operation with the users should have a positive attitude towards the working method. Problems can occur if the civil servant does not consider interaction useful. In his paper on participatory planning in Finland, Paldanius claims that many planners consider planning a matter for experts, not for interest groups (Paldanius 1997). He claims that the education and identity of these planners is a reason for this aversion to public participation. The training of planners is, in general, strongly based on the pursuit for objective scientific data. However, the current attitudes among civil servants seem to be more positive (personal opinion formed during several discussions and seminars on the subject). Communication with the resource users is seen to improve the quality of the regulation process.[3]

8.3.2 The Common-pool Theory

The common-pool theory, presented by Ostrom (1990) assumes that individuals do not act opportunistically in every case and thereby challenges, among other theories, the tragedy of the commons. Ostrom says that when aiming for outcomes other than what she refers to as 'remorseless tragedies' (The Tragedy of the Commons, The Prisoner's Dilemma and the Logic of Collective Action), it is important to 'enhance the capabilities of those involved to change the constraining rules of the game'. An application of this to today's situation could be that the users of a resource would be stimulated to communicate and organize the management of the resource, instead of the environmental administration informing the individual directly of rules and regulations, which disregard the community. The common-pool resource theory assumes that resource users communicate with each other and that they can agree on a common set of rules to be applied to the resource system. This theory, however, is not the optimal solution for each case, but it is relevant to the theory about co-management of nature resources because it suggests success factors for it. The concepts of the theory are defined in Box 8.1.

Box 8.1 Concepts of the common-pool resource theory (Dietz et al. 2002, Ostrom 1990)

Common - A variety of resources and facilities or property institutions with some kind of joint ownership or access
Common property - A common resource for which humans have arranged the management
Common pool - A resource from which it is difficult or impossible to exclude someone

---

[3] Personal observation from the seminar *Hyvät käytännöt vuorovaikutteisessa suunnittelussa* (Good practices in interactive environmental planning 2005).

Common pool resource - A valued natural or human-made resource or facility that is available to more than one person and can be subject to degradation if it is overused
Resource system - Stock of the resource. It is the basis of the resource that can be renewable or non-renewable
Resource unit - The actual good taken from the resource system
Appropriator - The user of the resource system
Provider - The individuals caring for the maintenance and supply of the resource system

The common-pool theory generally implies that the more influence the users have in management, the better the chances of success. Accordingly, it is important that the users are allowed to participate not only in defining the specific rules and regulations for using the resource, but also in defining the constitutional rules for how the management process should be constructed. The users should be included in monitoring the compliance of the regulations or the monitors should be accountable to the users. It should also be possible for the users to revise rules (See Box 8.2.).

Box 8.2 Principles increasing the performance of institutional design (Dolšak and Ostron 2003)

1. Rules are devised and managed by resource users.
2. Compliance with rules is easy to monitor.
3. Rules are enforceable.
4. Sanctions are graduated.
5. Adjudication is available at low cost.
6. Monitors and other officials are accountable to users.
7. Institutions to regulate a given common- pool resource may need to be devised at multiple levels.
8. Procedures exist for revising rules.

When the resource users have the chance to create and manage the rules for the use of the common-pool resource, the rules will be better applied to the local conditions and characteristics of the resource. Additionally, the users will be more familiar with the regulations and, for that reason, have a tendency to act in accordance with them (Acheson and Brewer 2003).

The characteristics of the appropriators and the choices made by them also have a decisive impact on the success of the management. Ostrom has concluded that the main factors influencing the appropriators' choices of strategy include expected benefits, expected costs, internal norms and discount rates (Ostrom 1990). The discount rates are affected by the range of opportunities an appropriator has outside the resource system.

Ostrom (1990) distinguishes three different management rules for a resource, operational, collective-choice and constitutional-choice rules. The most concrete level is formed by *operational rules*. Operational rules regulate the practical actions of the appropriators and day-to-day decisions, the withdrawal of resource units, provision, monitoring and enforcement. Operational choices are, to some extent, affected by the *collective-choice rules*. Collective-choice rules are used by the appropriators, officials and external authorities in making policies about how the resource should be managed. Finally, *constitutional-choice rules* determine who is eligible and how the collective-choice rules should be set. In a common-pool resource with open access, the appropriators have access to all rule levels.

When local actors and environmental authorities meet, the civil servants are in a peculiar situation. They establish a relationship with the local inhabitants, who often have a sceptical attitude towards environmental authorities and everything that even remotely reminds them of land use regulation. During the meeting a relationship is established. The civil servants use their social skills to establish trust in the environmental authorities. The civil servants representing the environmental authorities become a kind of mediator or messenger for the authorities. They must explain to the meeting's participants what protective measures are mandatory for biodiversity according to national legislation. However, a more important task is to explain why the protection is necessary and in a concrete way legitimize the regional, national, European and even global efforts to support biodiversity legitimacy.

8.4 The Case: Bergö-Malax' Outer Archipelago

Studying the case of Bergö-Malax' outer archipelago is a work in progress. The first observations on the archipelago were made in April 2005.[4] In addition to observations made at the meetings, four thematic interviews with the representatives of the environmental authorities involved in the planning process of Bergö-Malax' outer archipelago have been carried out. The study will continue with further observations and interviews with the local people who have been actively participating in the planning process.

8.4.1 Background

The Kvarken Council is a Nordic cross-border cooperation association for the Ostrobothnian counties in Finland as well as for the county of Västerbotten and the municipality of Örnsköldsvik in Sweden. The council works on both the Swedish

---

[4] Meetings were held on the 4th, 5th and 6th of April, on the 28th of November and on the 1st of December 2005.

and the Finnish side of Kvarken to increase co-operation between the nations. Together with the West Finland Regional Environment Centre and Metsähallitus, the Kvarken Council has developed a project called Kvarken Environment, which contains the subproject Grön Bro II (=Green Bridge II). The intention of Grön Bro II is to prepare Kvarken to become a World Heritage Site in 2006. The project develops the infrastructure of the nature resources in the region in accordance with the requirements for world heritage status. The Kvarken region has large Natura 2000 areas; therefore, the work carried out there is closely related to the Natura 2000 conservation programme. Efforts are being made to investigate, plan and put into practice the management and use of the nature protection areas in the region. Putting together a management and utilization plan for the Bergö-Malax' outer archipelago is one of the concrete projects initiated by the regional environmental authorities and supported in terms of finance and expertise by the Grön Bro II project. The project is a possible 'good example' of a type of co-management because of the outside financing of the Kvarken Council and EU Interreg. Without the external resources, it would probably not be possible to carry out the project using co-management as the working method.

8.4.2 Description of the Area

The Bergö-Malax' outer archipelago is part of the village of Bergö, which is the closest inhabited location. Bergö village has approximately 500 inhabitants, who primarily support themselves by fishing and navigation. In 2003, the village had approximately 40 full-time fishermen. Until 1973, Bergö was an independent municipality. Today, it belongs to the Malax municipality, but Bergö's independent history still plays an important role in the community's spirit. The inhabitants of Bergö have a very strong connection to the archipelago outside their village, where the study area called Bergö-Malax' outer archipelago is located. If anyone can be considered the inhabitants of the area, they can.

The study area is part of the Natura 2000 network. It consists of several islands with some seasonal habitation. The entire Kvarken area is characterized by a fast land up-lift, which is a result of the ice age. Moreover, the up-lifted areas of land form a succession biotope, which is biologically very interesting. The up-lifted lands have also become property of the village commons. In Bergö-Malax' outer archipelago, the Bergö village common owns a large part of the seashore areas. The area is regularly used for fishing, hunting and recreation. Those using the area are individuals and associations that come from the surrounding area, but also from further away. When it comes to managing the area, conflicts have occurred between the local users of the resource and the environmental authorities. Because the Natura 2000 programme regulates the use of the area in a way that interferes with its use, the users feel that the possibilities for fishing, hunting and recreation are threatened. Outside recreational users also put pressure on the area. This creates some conflict, particularly between professional and recreational fishermen. Some professional fishermen depend on this area for their livelihood.

Bergö-Malax' outer archipelago is primarily owned by the village common of Bergö. Some parts of the land are owned by the state. There are protection measures being carried out on the archipelago; the landowners can choose between protecting the land that is in their possession and selling the land area to the state. This means that the proportion of state owned land might increase in the future. Furthermore, the water areas are both privately and state owned. Some parts of the state owned land are public whereas other parts are administered by Metsähallitus, which means that utilization is restricted (correspondence with civil servant at the environmental authorities, 1.11.2005).

8.4.3 The Management and Utilization Plan

The process of creating the plan for Bergö-Malax' outer archipelago is carried out in public meetings. The following stakeholders were invited to and attended the first planning meeting (protocol from the meeting 8.11.2004):

Local associations for fishery
Local associations for ancient monuments
Local associations for hunting
The board of local community landowners
The Bergö island council (Bergö is the closest inhabited village)
The municipality Malax
Seasonal inhabitants
The Swedish Ostrobothnian hunting district
The regional environment centre

It was clearly stated by the authorities that participation should not be limited and that the participant contact list could be added to at any time. The first meeting took place in November 2004, during which the project of putting together the management and utilization plan was introduced. After the initial meeting, all subsequent meetings have had themes. The themes have been hunting, fishing, recreation and history. The intention is that the participants choose a group and attend the series of meetings of this particular group. The theme groups are intended to work in parallel and the plan is scheduled to be finished in 2007.

8.4.4 Bergö-Malax' Outer Archipelago in Terms of Co-management and the Common-pool Theory

Coming back to the division of co-management into the four sub-categories illustrated in Fig. 8.1 (Carlsson and Berkes 2003), the situation in Bergö-Malax outer archipelago can best be illustrated by Figs. 8.1a) and 8.1b). Traditionally, communication between the environmental authorities and the local inhabitants has been one or two-directional information. Where co-management has occurred, it has been mainly the type illustrated by Figure 8.1a), i.e. two separate spheres, the private and the state, have communicated and exchanged information. The participatory process in Bergö-Malax outer archipelago is an attempt to create an arena where the two spheres can overlap, as illustrated by Fig. 8.1b). Figure 8.1b) implies that the state and private spheres partially overlap. In the planning process, the authorities try to create a chance for representatives of the state and the private spheres to establish cooperation, so that the two spheres are not so strictly separated. An example of this is the co-writing of the draft of the plan. Meeting participants compose a concrete draft, which will eventually become the actual management and utilization plan, using a computer together with a projector. Consequently, everyone can equally see and influence what is being put into the draft. The draft is then sent to the environmental authorities for approval.

The illustration presented by Carlsson and Berkes (2003) is problematic because it does not consider the difference between the private sphere of individuals as actors and the part of the private sphere where individuals are part of a community. The situation where the individuals act as owners of commons, is characterized by the group of individuals being formed independently of the state's intervention. In Bergö-Malax' outer archipelago, social structures improve the chances of establishing management cooperation. Many individuals are part of associations that have already managed common properties before the environmental authorities started showing an interest in protecting the area. Co-management as cooperation between the environmental authorities and the commons differs from cases where the authorities cooperate with individuals.

As Borrini-Feyerabend et al. (2003) and Jentoft et al. (1998) argue, there have often been structures of co-management in the past. This is also true in Bergö-Malax outer archipelago. Important forms of co-management of the resources already existed before the environmental authorities started showing an interest in the area. The local fishing and hunting societies have practices for putting restrictions on the appropriation of the resource. For example, every year these local societies decide on regulations for the equipment that is used, the time when hunting in certain areas is allowed, and so forth.

The archipelago's resource users are primarily people from the village of Bergö. However, people from the region also frequently use the area, for example, sailing clubs. In addition, occasional visitors from farther away also use the resource, mainly for recreation. The providers are the local societies that organise regulated fishing and hunting as well as Metsähallitus and The West Finland Regional Environment Centre, which try to direct the recreational users of the area into sites less sensitive to human use. The West Finland Regional Environment Centre, Metsähallitus and the Kvarken Council can also be seen as providers because they provide the institutions for co-management through arranging public meetings to gather together local users.

From the local appropriators' point of view, the area's main resources are fish for professional and recreational fishing and seabirds for recreational hunting. At the meetings, issues concerning these resources cause the most intense debates.

Another resource valued by the local user is the possibility to use the area for recreational purposes, both for seasonal habitation and occasional visits. At the meetings the participants are often concerned about the limited right to build new summer cottages and about regulating the activities around the existing summer cottages (e.g. cutting wood and keeping up fairways by dredging). The environmental authorities regard biotopes of natural land up-lift and fladas in natural succession to be particularly valuable (see Box 8.3). They also see the seabirds as a valuable resource in their aim to protect biodiversity, especially the greater scaup (*Aythya marila*), which is considered to be of particular value because it is very rare in Europe.

The area's resources can roughly be divided into two categories: biotopes and animal species. The biotopes are valuable from both a recreational and biodiversity aspect. Recreants appreciate the unspoiled and beautiful nature whereas scientists appreciate the area's biodiversity. The animal species tend to be valued as prey and as part of a general perception of a living nature for the local users in the area. The environmental authorities see the species as a part of biodiversity; the more rare the species from a European or global point of view, the more important it is to protect it. The environmental authorities aim at keeping the biotopes as natural and pristine as possible, whereas the users of the resource want to guarantee the right to use the areas for building sites.

In many cases, the environmental authorities and the local users share the perception of what measures are needed to sustain the resources. Although the local users want to practice hunting and fishing, they depend on stable fish and prey stocks. It is thus in their interest to preserve good living conditions for fish and prey. Local users also often want the area to preserve its recreational value, which means that building activities should stay limited, although allowed.

Box 8.3 The controversial value of the flada

The Flada and its controversial value

An example of different views on what is valuable is the *fladas*. A *flada* is a bay that is separated from the sea, due to the rising of land or other geo-morphological functions, and will eventually become a separate lake or glosjö. In the Kvarken region, there are several successions from flada to glosjö, which, according to the environmental authorities, have a high biological value. On the other hand, the people who have their summer cottages on the shore of the flada are likely to consider accessibility by boat to the summer cottage valuable. A dispute arises between those who want to safeguard the natural succession of the flada and the inhabitants of the summer cottages who want to keep the flada open by dredging. In the case of Bergö-Malax' outer archipelago, the parties came to a compromise where the cottage owner is allowed to dredge as long as he/she follows acceptable practices that do not harm the biological features of the flada.

Both the co-management and the common-pool theory emphasize the importance of the government supporting existing structures in the community rather than trying to replace them with new ones. It is important that the rules for sustaining the resource are revised and managed by its users. The monitors and other officials should also be accountable to the users (see Box 8.2, Dolšak and Ostron 2003). Jentoft et al. (1998) claim that community cohesion is important for resources management. In general users tend to act more opportunistically as individuals than as part of a community. Traditionally, the Finnish environmental authorities have usually communicated quite directly with the individual about possible protection of land areas or state purchase of privately owned land. During the process of creating the management and utilization plan in Bergö-Malax outer archipelago, the community played an important role. Initially, the local associations, who had considerable interest in the area, e.g. fishing and hunting, and landowners' associations were invited to participate. This action indicates that there was a community-based approach in the process. Authorities communicating on the local level turned to the associations and not directly to individuals.

The three levels of management presented by Ostrom (1990) illustrate how decision-making is carried out and how it affects the use of the resource. Ostrom describes an open access common-pool resource as a resource in which the appropriators have access to all rule levels. She also describes a situation where the users only have access to operational rules and have to obey the collective-choice and constitutional-choice rules that are established without their consent. In the case of the Bergö-Malax outer archipelago, the users mainly have access to the operational rules. When it comes to collective-choice rules as well as constitutional rules, it looks quite different, however. To a large extent, it is the authorities that decide not only on the management policies of the resources, but also on the definition of who should have a say on how the policies are carried out. The planning process is a way of giving the users insight into and understanding of the collective-choice rule making, as well as why the rules exist and how they should be obeyed. There might be a slight misunderstanding between the participants and the representatives of the environmental authorities. It is easy to get the impression that the authorities also want to give the local users influence on the collective-choice level. However, during the interviews, representatives of the environmental authorities emphasized that the participation process is primarily an action to increase the understanding of nature protection among the locals.

During the process of creating the management and utilization plan, the participants can indeed influence the process at some points. They can, for example, add to the list of which parties should be invited to the meetings. The participatory process is, however, presented to the participants as an opportunity rather than something to evaluate critically. The participants are not encouraged to take a stand on the way the process is carried out.

The open meeting is an important event in the process of co-management. In the case of a conflict, it is very useful to bring different stakeholders together. When the stakeholders can communicate directly with each other, there is a good chance that mutual understanding, or at least acceptance of the fact that there are different opinions, will occur. Turunen presents five factors influencing the success of the meeting: (1) the efforts for interaction, (2) the technical arrangements, (3) the idea behind the event, (4) how the event is connected to the actual planning process, and (5) communication. He suggests two alternative aims of the meeting: to avoid problems and to give the result or the organisation more value. There might be occasions when the meeting for the affected parties is arranged in order to avoid objections at a later stage. The meeting is held only to fulfil legal requirements. There are, however, cases where openness during the planning process is highly valued. The final result is then seen to be of better quality because there has been interaction and a broader range of interests has been taken into consideration. It is important to consider the connection between the meeting and the planning process and to make sure that what is discussed during the meeting will influence the final product (Turunen 2005). If the participants of the meeting are not shown that their opinions influence the process, this can engender feelings of frustration.

At the meetings in Bergö, different types of people participate. Based on our observations, we divide the participants into four categories: landowners, users, local visitors and distant visitors. During the meetings, each participant type has a different authority (opinions weigh differently depending on the group) and each type has a differing perception of its right to have an opinion on a particular matter. Landowners usually have a strong feeling of being affected, partly due to the strong landowner tradition. Their perception of security is threatened by regulations on the use of the area. In the users group, we see fishermen and hunters who, to a varying degree, depend on the concrete resources of the area. The local visitors are people who mainly use the area for recreational purposes, however, because they live in the vicinity of the area or they have a strong connection to it through, for example, relatives, they are accepted in the community. Distant visitors use the area for recreational purposes. They have no social connection to this particular area. Hence, this group has the least authority at the meetings.

The meetings have a very open and communicative character. The knowledge of the local participants is used to make the management and utilization plan compatible with the requirements of the area. The additional studies that are done in order to complement the data for the area are carried out openly and with the local users being aware of them.

According to our observations so far, strong management regimes with a long history exist in the area. When the environmental authorities make an effort to establish co-management of the area, these regimes will have an important role in building up the potential for success.

8.5 Conclusion

An active discussion can be observed among Finnish environmental administrators about the increasing role of deliberative methods when putting environmental policy into practice. The need to enhance legitimacy and local anchoring in nature conservation is well acknowledged, especially considering the lessons learned during the initial stage of the implementation of Natura 2000. In the case study of Bergö-Malax' outer archipelago, we observe top–down conservation policy being alternated with an effort to establish a type of co-management of a nature resource.

It has been realized that being innovative at the local level concerning new models of nature resource management is a challenge to environmental policy being regulated on the EU and national levels. The mandate of the civil servant who is involved in the participatory process from his/her organization is of great importance. The civil servant has to receive enough responsibility to be able to step out of the role of strict nature conservationist and see a larger variety of factors that depend on and influence the management of the nature resource. The freedom of interpretation in nature conservation regulations might pose a problem in establishing trust at the local level because what the civil servant finds legally acceptable might not be accepted at higher levels in the organization. Therefore, it can be problematic for the civil servant to establish trust in the environmental authorities locally. Co-management might be a possibility to reintroduce the undermined aspects of sustainability, namely the social and economic aspect. It is a pre-condition for co-management that the civil servant has the mandate of his/her organization to carry out the planning with sufficient authority. Otherwise, it is our contention that communication between the locals and the authorities will only have an informative function.

In Bergö-Malax outer archipelago, important systems of self-regulation have been observed. These will play an important role in the success of co-management. When the authorities make an attempt to establish co-management in the area, it is important that local regulatory systems are taken into account. It is also important to put the emphasis on discovering and supporting old structures instead of starting from scratch to create new institutions.

A possible product of co-management is vitalization of the community. In the discussion on the profitability of using participative methods, vitalization should be taken into consideration. The social improvements in the community that emerge when the inhabitants come together to deliberate over the future of their surroundings should not be undervalued. At another location in the Kvarken region, a community gained a great deal from participative actions because common ground was found by the inhabitants and new social structures were formed.

It is of importance for the success of co-management that the users have influence at the collective-choice level. As the common-pool theory suggests, the lack of influence both on the process of choosing the methods for co-management and on the monitoring implies that co-management might possibly fail. In the case of creating a management and utilization plan for Bergö-Malax' outer archipelago, the participants have some influence on how the planning and management are carried out. It is, however, important to see the whole picture. The creation of the management and utilization plan is a result of a protective regulation, Natura 2000, which has already established a framework for the utilization of the area. The process of implementing Natura 2000 did not include consultation with the local actors. According to the common-pool theory, it is important that the local users are included not only when following existing rules, but also in defining the rules. As a separate process, the work with the management and utilization plan can be seen as successful and legitimacy-enhancing. However, looking at the whole picture, it is easy to see the element of an attempt to increase acceptance of a conservation programme that was enforced using strong top–down means.

Acknowledgments This chapter is based on research from the projects *Management cultures and the implementation of supranational environmental agreements – The case of Natura 2000 in Finland* (2001–2004) and *New models for policy and decision making in nature resource management – multiple knowledge use and local/community participation in nature resource regulation and protection* (NERM) (2005–2007). Both projects are financed by the Helsinki Environmental Research Centre at the University of Helsinki. Important contributions in the form of discussion and comments have been received from Erland Eklund and Ville Klemets, members of the project NERM.

References

Acheson, J.M. and J.F. Brewer (2003). 'Changes in the Territorial System of the Maine Lobster Industry.' In N. Dol²ak and E. Ostrom, eds. *The Commons in the New Millennium. Challenges and Adaptations.* Cambridge: The MIT Press.

Bonney, L. (2004). 'EU:s implementering av Natura 2000 – interaktionen mellan EU och den finländska nationella nivån.' *SKKH Reports and Discussion Papers.*

Borrini-Feyerabend, G., M.T. Farvar, J.C. Nguinguiri and V. Ndangang (2000). *Co-management of Natural Resources: Organizing Negotiation and Learning by Doing.* Heidelberg: Kasparek Verlag.

Carlsson, L. and F. Berkes (2003). 'Co-management Across Levels of Organization: Concepts and Methodological Implications.' Lead paper prepared for the Resilience panel at the Regional Workshop of The International Association for the Study of Common Property (IASCP) "Politics of the Commons: Articulating Development and Strengthening Local Practices". Thailand.

Dietz, T., N. Dol²ak, E. Ostrom and P.C. Stern (2002). 'The Drama of the Commons.' Introduction in E. Ostrom, T. Dietz, N. Dol²ak, P.C. Stern, S. Stonich and E.U. Weber, eds. *The Drama of the Commons.* Washington, DC: National Academy Press.

Dolšak, N. and E. Ostrom (2003). 'The Challenges of the Commons'. In N. Dol²ak and E. Ostrom, eds. *The Commons in the New Millennium. Challenges and Adaptations.* Cambridge: The MIT Press.

Hardin, G. (1968). 'The Tragedy of the Commons'. *Science* 162: 1243–1248 (www.sciencemag.org).

IUCN (1997). Resolutions and Recommendations, World Conservation Congress, Montreal, Canada, 13–23 October 1996 (http://iucn.org/wcc/resolutions/resrecen.pdf, 5.1.2006).

Jentoft, S., B. McCay and D.C. Wilson (1998). 'Social theory and fisheries management'. *Marine Policy* 22(4–5): 423–436.

Jentoft, S. (2000). 'The community: a missing link of fisheries management.' *Marine Policy* 24: 53–59.

Jokinen, P. (1995). 'The development of agricultural pollution control in Finland.' *Sociologia Ruralis* 35(2): 206–227.

Korhonen, V. (2004). 'Konflikten kring implementeringen av Natura 2000 i Finland – en konflikt mellan olika kunskapssystem samt mötesplatsernas betydelse för skapandet av tillit.' *SKKH Reports and Discussion Papers.*

Malmsten, A. (2004). 'Natura 2000-verkosto maanomistajien silmin Lounais-Suomessa' (The Natura 2000 network through the eyes of the landowners in Southwest Finland). *Terra* 116(2): 67–76.

Nieminen, M. (2000). 'The Incontestable Nature of Biodiversity.' Paper for the POSTI conference on Policy Agendas for Sustainable Technological Innovation, London.

Ostrom, E. (1990). *Governing the Commons: The Evolution of Institutions for Collective Action.* Cambridge: Cambridge University Press.

Paldanius, J. (1997). 'Experiences of Public Participation in Finland'. In B. Sohlberg and S. Miina, eds. *Conflict Management and Public Participation in Land Management.* EFI Proceedings No. 14.

Turunen, J.-P. (2005). 'Vuorovaikutuksen järjestäminen kaavaprocessissa.' Presentation at the seminar *Hyvät Käytännöt vuorovaikutteisessa suunnittelussa* arranged by the Finnish Environment Institute 10–11.10.2005.

Chapter 9
Visions and Scales of Nature and Society in Nature Management

Jac A. A. Swart

Based on the case described by Susanna Björkell in this volume, this commentary explores the role of visions of nature and scales in relation to environmental and nature management conflicts. It argues that such conflicts are characterized by multiple visions of nature related to spatial and temporal scales. The involvement of such scales in nature management policy is related to contrasting knowledge traditions ranging from practical and local to scientific approaches and also implies a complex relationship with different political scales, whose jurisdictions range from local to national or supranational. As a result, scale discordance can be related to political scale jurisdiction conflicts and the issue of governance generally. The involvement of multiple scales not only increases the complexity of environmental problems, but may also provide opportunities for new approaches to these problems. A pragmatic approach is advocated, which suggests exploring different strategies and the conditions under which these strategies may work. Grasping such opportunities requires the recognition of different scales, a respectful treatment of different views, open forums for deliberations, the assessment of information related to all stakeholders, and fair decision-making procedures.

9.1 Introduction

Susanna Björkell describes how the implementation of the EU Natura 2000 Network in Finland confronted the existing Finnish approach to nature protection, especially its strong relationship to landowner traditions. The top–down EU approach to nature protection intensified an already existing aversion to the European Union within the Finnish rural community, which had been struggling to cope with the competitive EU market policy. The implementation of the Natura 2000 Network stagnated due to this confrontation. As a result, Finnish environmental authorities have become much

Jac A.A. Swart
Science & Society Group (SSG), Faculty of Mathematics and Natural Sciences,
University of Groningen, Netherlands
E-mail: j.a.a.swart@rug.nl

J. Keulartz and G. Leistra (eds.), *Legitimacy in European Nature Conservation Policy:*
Case Studies in Multilevel Governance, 127–135. © Springer 2008

more focused on co-management approaches. Björkell analyses a particular example of the current management approach - the Bergö-Malax outer archipelago lying off the west coast of Finland - using theoretical insights drawn from co-management and common-pool resource theory.

Although there are many definitions and methods of co-management, a common characteristic is that users of a common-pool resource have significant rights of accessibility but also responsibilities, in order to prevent the occurrence of the 'tragedy of the commons' as described by Hardin (1968). Co-management can thus be seen to mirror the institutional arrangements in which management responsibilities are shared between users and the government (Plummer and Fitzgibbon 2004). Carlsson and Berkes (2003) consider co-management to imply systems of governance in the sense that it relies on the societal coordination of societal systems. While such systems may also be possible without government involvement, co-management can be seen to entail a shift towards governance.

This shift towards governance and thus towards societal and political structures and conditions with respect to nature management may, however, overshadow ecological perspectives in nature management and conservation. Nature management conflicts do not need to be solely based on governance issues, such as those seemingly suggested by co-management and common-pool resource theories, as described by Björkell, but may also be initiated by conflict concerning the desired conditions and type of natural system, and related to this, the desired role of human beings with respect to nature (Van der Windt et al. 2006). Social conflicts may arise, for example, by forcing a particular and distinctive perception of nature upon stakeholders who subsequently question the legitimacy of this view (see also Korthals in this volume, pp. 249–253). If these perceptions of nature were more consistent with local perceptions, their enforcement might not be questioned and may even be perceived as supporting the local tradition. Therefore, we also need to analyse nature management conflicts in terms of the stakeholders' visions or perceptions of nature and to explain why and how such conflicts can become governance conflicts. Running the risk of being overly speculative, I will endeavour to provide an outline of such an analysis in this commentary.

9.2 Visions of Nature: Landowners Versus the EU Ecological Network

In previous publications my colleagues and I have proposed a model for the valuation of nature in which three main perspectives or visions of nature are identified, i.e. the functional, the arcadian and the wilderness visions, which are each based on underlying ecological, ethical and aesthetic considerations (Swart et al. 2001; Keulartz et al. 2004; Van der Windt et al. 2006). The functional vision considers nature primarily in terms of its value to humanity, focusing on such aspects as timber production, fishery yields and opportunities for recreation. This is the rationale behind the conservation and restoration of many accessible reserves, parks, gardens and utilized meadows, and also behind many activities affecting the landscape such as agriculture and fishing. In short, the functional vision is dominated by utilitarian ethical considerations, with aesthetic appreciation often being based on a functional and formal view of aesthetics. The most important scientific approaches within this tradition are production ecology and population dynamics.

In contrast, proponents of the wilderness vision particularly value nature in its unspoiled state, characterized by authentic species and natural constitutive processes such as erosion and sedimentation, dispersion, and predation. The utilization of nature is only acceptable if the effects are negligible. Ethical considerations often refer to the intrinsic value of plant and animal species, ecosystems, and ecological processes. According to this perception, aesthetic appreciation is based on the value of the authentic wilderness, possibly revealed by scientific research. The main scientific approach is ecosystem ecology, which focuses on ecological processes.

The arcadian vision takes more of an intermediary position. It focuses on semi-natural and extensively used cultural landscapes where human impact is considered to be positive when it contributes to biodiversity and the landscape. Its proponents stress the cooperation between humanity and nature and often refer to historical elements, traditional knowledge of land management, and both extensive and small-scale utilization practices. The aesthetic appreciation of arcadian nature or landscapes is often influenced by historically based, subjective considerations founded on community ecology. The ethical dimension of the arcadian perspective is characterized by ideas such as stewardship, care and the extent to which humans are connected with the biotic and non-biotic environment.

These three distinct visions of nature form an ideal-typical model. In practice we often find intermediate positions, as is demonstrated by Fig. 9.1 (based on Van der Windt et al. 2006, Fig. 1). This figure shows that conservationists, conservation officials and farmers often hold different combinations of the three visions of nature distinguished in this model. Farmers are generally functional arcadians, whereas con-

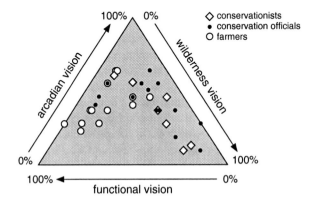

Fig. 9.1 Distribution of visions of nature among 34 individual actors involved in nature conservation based on a multiple-choice survey (overlaping data points have identical positions). Each point represents the percentages of wilderness, functional and arcadian answers to nine survey questions (see also Van der Windt et al. 2006). In general we see a distribution over two sides of the triangular vision-space, with a bridging role for the arcadian vision

servationists and conservation officials lean towards a wilderness-arcadian stance. The model is also relevant to issues concerning sustainability, including socio-economic issues, and the question of the extent to which nature should be considered vulnerable or resilient to human impact. Adopting the approaches of Dobson (1998) and Swart et al. (2001), a model on visions of sustainability was developed in parallel with the model of the visions of nature (Swart and van der Windt 2005).

In the Bergö-Malax case described by Björkell, we can distinguish similar positions on nature and sustainability to those distinguished by Swart and co-workers, described above. The landowners' traditional view of nature exhibits a mix of functional and arcadian visions, which find present-day local land use, historical traditions, and the use of traditional local knowledge to be important. Alternatively, we can see a mix of wilderness and arcadian visions in the Natura 2000 Network approach. The EU habitat directive, in which the Nature 2000 Network was announced, aims to 'contribute towards ensuring bio-diversity through the conservation of natural habitats and of wild fauna and flora' (Article 2.1). According to Article 2 of the habitat directive, natural habitats include both natural and semi-natural areas (Office for Official Publications of the European Communities 2004). The network approach is in accordance with the wilderness perspective, in so far as it stresses the importance of setting aside large areas for natural processes, while the reference to semi-natural areas recognizes the arcadian nature conservation tradition on the European continent.

9.3 Scale Discordance

Thus, in spite of the converging role of the arcadian vision, in practice two positions come to the fore, the wilderness-arcadian position, related to the ecological network approach and a more functional-arcadian position related to the landowners' approach. In practice these two positions are also related to spatial and temporal scales. The network approach, stressing the importance of natural processes such as migration, dispersion of wild species, sedimentation, erosion and evolution, as important constituent elements of nature, often implies large geographical and temporal scales. In contrast, the landowners' approach, stressing human utilization, focuses much more on local and short-term (often annual) scales.

As such, we can distinguish two dimensions in nature management conflicts: on the one hand there is a conflict in visions of nature ranging from functional to wilderness perceptions, and on the other hand, but interrelated, there is a difference in scale ranging from local and short-term to global and long-term perspectives.

Applying this dual conflict model to the Bergö-Malax case, we can conclude that the network perspective on nature management is conditioned by ecological considerations strongly related to fluvial or continental biogeographical and long-term scales, while the landowners' perspective focuses on local and short-term scales (see Fig. 9.2).

Understanding the complexity of ecosystems requires the recognition of the effects of spatial and temporal scaling. In ecological sciences, the concepts of

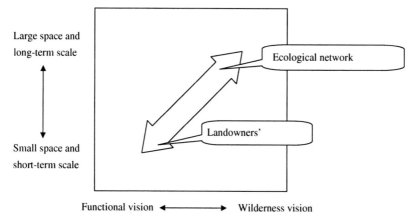

Fig. 9.2 A hypothesized relationship between visions of nature and the scales involved. The ecological network approach expresses a wilderness vision of nature and larger spatial and temporal scales. The landowners' approach focuses much more on a functional vision of nature and smaller scales

temporal and spatial scales are increasingly recognized as fundamental terms (Schneider 2001; Levin 1992; Wiens 1989). Many ecological phenomena such as extinction, distribution and disturbance are related to scale levels affecting the observation and predictability of ecological phenomena. Moving from lower or higher spatial or temporal scale levels may lead to the appearance or disappearance of such phenomena. On a local scale level a species may be abundant, whereas on a global scale level such a species may be on the verge of extinction. In addition, processes on higher scale levels often occur much more slowly than those on lower scale levels. However, processes and structures on different scale levels are causally related through complex, sometimes worldwide phenomena such as food webs, migratory patterns of species, prey-predator relationships and physical circulation systems, such as wind, rivers or ocean currents. As a consequence, problem definitions, observations and proposed solutions with respect to environmental and ecological issues do not need to occur on the same scale level.

The recognition of scales in relation to visions of nature in ecology has implications for the analysis of nature conservation and management conflicts. Firstly, as visions, these different scales are also related to different knowledge traditions. Scientific ecological analysis and assessment often focuses on larger scales with universalistic claims, whereas practical management mainly occurs on the local scale and is undertaken by stakeholders and local decision-makers, whose vision of nature relies partly on practical knowledge and traditions directly related to the local situation (Cash and Moser 2000; Wilbanks 2002; Berkes 2006).

Secondly, scale phenomena are not only relevant to ecological and environmental sciences but have implications in social, economic and political terms. They imply a complex relationship of different political jurisdictions, ranging from local to national or supranational, as is shown by Björkell in the discussion of the role of the

Table 9.1 Two contrasting conservation and management approaches imply besides different visions of nature different ecological scales with, knowledge traditions and levels of decision-making

Conservation and management approaches	Visions of nature	Ecological scales	Knowledge traditions	Levels of decision-making
Bottom–up (Landowners' approach)	Functional/ arcadian vision	Small spatial and short-term scale	Practical and local knowledge claims	Local decision-making
Top–down (Network approach)	Wilderness/ arcadian vision	Large spatial and long-term scale	Scientific and universalistic knowledge claims	(Supra) national decision-making

EU habitat directive in EU nature management policy. Table 9.1 demonstrates the complex scaling aspects in conservation and management.

Scales in ecology and conservation are thus related to political jurisdiction scales. As an example, with reference to the different management rules examined by Ostrom (1990), Björkell notes that stakeholders should not only be involved in applying rules but also in defining and designing them. Changing the rules in the current political environment needs the approval of higher political scale levels, perhaps as high as EU level, as a consequence of the habitat directive. Nonetheless, the Bergö-Malax case demonstrates a reversed dependency, with the EU top–down approach failing through insufficient awareness of the role demanded by the local community.

Understanding and taking into account these different scales is thus a prerequisite for successful nature conservation and management. However, Cash and Moser (2000) conclude that with respect to environmental problems, science and policy-making often only pay attention to one of the relevant scales, missing cross-scale interactions or scale discordance. In the Bergö-Malax case, this not only seems to be the problem with the EU approach, but also with the co-management approach. Its emphasis on local resources may comply rather well with the functional-arcadian position of nature, but still ignores the ecological insights that are manifest on larger biogeographical and temporal scales.

Ecological and societal systems are not only complex and cross-scaled, they are also increasingly intertwined in synergetic dependency relationships (Rosenzweig 2003). Many ecosystems are 'humanized' as a matter of fact and have become dependent on human regulation for their persistence. Indeed, as is stated by Jentoft (2000: 53) and cited by Björkell, 'viable fishery stocks require viable communities'. This insight is not just a prerequisite for successful co-management on a local scale level, but also at higher scale levels. Because of the multiple scale character of environmental problems, a respectful relationship between political jurisdictions is needed. To put it in general terms, sustainable ecosystems require a decent society.

Thus, assessing and managing complex and multi-scaled environmental and ecological systems requires an integrated multi-level and multi-scaled approach. Since most expertise is related to a particular scale and often to a particular level within a scale, there is a strong need to integrate and coordinate the multiplicity of types of expertise involved. In academic research this recognition has led to the concept of boundaries. These boundaries lie between politics and science and are usually contested and negotiated by both scientists and politicians when science is brought into the political arena. Gieryn (1995) has labelled activities in which actors redefine what counts as being scientific or political, as 'boundary work'. In addition, other scholars have coined the term 'boundary organization' for institutions that mediate between politics and science. Such institutions facilitate the involvement of different actors in boundary work and integrate contrasting rationales, deriving their own legitimization and stability from this intermediary position (Guston 2000). Cash and Moser (2000) argue that in environmental affairs these boundary organizations not only mediate between scientists and decision-makers but also between these actors operating on different spatial scale levels.

Thus, boundary work on environmental issues involves different paradigms and different scales. However, one may still ask whether these institutions can really tackle this degree of complexity. By relying on boundary organizations we not only create additional institutions with their own rationales and interests, thereby increasing the complexity of the field, we also create multiple interfaces. In controversies, the expertise of intermediary boundary organizations may therefore also be contested.

9.4 Conclusion

We may conclude that the involvement of scales increases the complexity of environmental and ecological issues. In addition, social and ecological systems are often connected through these various scales (see also Berkes 2006). Recognizing the role of scales is important as it can suggest new directions for dealing with such environmental issues. Problem definitions, observations and proposed solutions arising on other scales and other levels may appear to be relevant. Overemphasizing the role of one level or solely using bottom–up or top–down approaches may be counterproductive, as sustainable management is founded on recognizing the links between different scales of concern (Wilbanks 2002). In my opinion we should strive for a pragmatic approach. This means the exploration of different strategies and the conditions under which such strategies may work. Bottom–up approaches as advocated by co-management and common-pool resource theories are important, but in the current trend towards globalization we also need global views and constraints, which can direct local action. More generally, we should downscale nature conservation and management when possible and upscale if required. Recognizing the multiple scale character of the current environmental and ecological crisis may generate solutions satisfying the conditions on different levels within different scales, with zone management (Weinstein et al. 2006/2007) perhaps being an example of such an approach. Important to this pragmatic approach is the respectful treatment of different views, open forums for deliberations, the assessment of information relevant to all stakeholders, and fair decision-making procedures. These conditions are related to a much wider raft of conditions such as high standards of education and a democratic and fair distribution of welfare within the current global society. Indeed, we need a decent society for a sustainable world.

References

Berkes, F. (2006). 'From community-based resource management to complex systems: The scale issue and marine commons'. *Ecology and Society*, 11(1): 45 [online] URL: http//www.ecologyandsociety.org/vol11/iss1/art45/.

Carlsson, L., and F. Berkes (2003). 'Co-management across levels of organization: Concepts and methodological implications.' Lead paper prepared for the Resilience panel at the Regional Workshop of The International Association for the Study of Common Property (IASCP), "Politics of the Commons: Articulating Development and Strengthening Local Practices", Chiang Mai, Thailand, July 11–14, 2003. http://eprints2.dlib.indiana.edu/archive/00001133/00/Lars_Carlsson.pdf, assessed on 6 May 2006.

Cash, W.D., and S.C. Moser (2000). 'Linking global and local scales: Designing dynamic assessment and management processes'. *Global Environmental Change* 10: 109–120.

Dolšak, N. and E. Ostrom (2003). 'The challenges of the commons'. In N. Dolšak and E. Ostrom, eds. *The Commons in the New Millennium. Challenges and Adaptations*, Cambridge: The MIT Press.

Gieryn, T.F. (1995). 'Boundaries of science.' In S. Jasanoff, G.E. Markle, J.C. Petersen and T. Pinch, eds. *Handbook of Science and Technology Studies*. Thousands Oaks: Sage.

Guston, D.H. (2000). Between Politics and Science: Assuring the Integrity and Productivity of Research. Cambridge: Cambridge University Press.

Hardin, G. (1968). 'The tragedy of the commons.' *Science* 162: 1243–1248.

Jentoft, S. (2000). 'The community: A missing link of fisheries management.' *Marine Policy* 24: 53–59.

Keulartz, J., H.J. van der Windt and J. Swart (2004), 'The case of Dutch nature policy. Concepts of nature as communicative devices.' *Environmental Values* 13: 81–99.

Levin, S.A. (1992). 'The problem of pattern and scale in ecology'. The Robert H. MacArthur Award Lecture, presented August 1989, Toronto, Ontario, Canada. *Ecology*, 73(6): 1943–1967.

Office for Official Publications of the European Communities (2004). Council Directive 92/43/EEC of 21 May 1992 on the conservation of natural habitats and of wild fauna and flora. http://europa.eu.int/eur-lex/en/consleg/pdf/1992/en_1992L0043_do_001.pdf, assessed on 19 May 006.

Ostrom, E. (1990). Governing the Commons: The Evolution of Institutions for Collective Action. Cambridge: Cambridge University Press.

Plummer, R., and J. Fitzgibbon (2004). 'Co-management of natural resources: A proposed framework.' *Environmental Management* 33(6): 876–885.

Rosenzweig, M.L. (2003). Win–Win Ecology. How the Earth's Species Can Survive in the Midst of Human Enterprise. New York: Oxford University Press.

Schneider, D.C. (2001). 'The rise of the concept of scale in ecology.' *BioScience* 52(7): 545–553.

Swart, J.A.A., and H.J. van der Windt (2005). 'Visions of nature and environmental sustainability: Shellfish fishing in the Dutch Wadden Sea.' *Restoration Ecology* 13: 183–192.

Swart, J.A.A., H.J. van der Windt and J. Keulartz (2001). 'Valuation of nature in conservation and restoration.' *Restoration Ecology* 9: 230–238.

Wiens, J.A. (1989), 'Spatial scaling in ecology', *Functional Ecology*, 3: 385–397.

Wilbanks, L. (2002). 'Geographic scaling issues in integrated assessments of climate change.' *Integrated Assessment* 3(2–3): 100–114.

Weinstein M.P., R.C. Baird, D.O. Conover, M. Gross, J. Keulartz, D.K. Loomis, Z. Naveh, S.B. Peterson, D.J. Reed, E. Roe, R.L. Swanson, J.A.A. Swart, J.M. Teal, R.E. Turner and H.J. van der Windt (2006/7) 'Managing coastal resources in the 21st century.' *Frontiers in Ecology and the Environment*, 5(1): 43–48.

Windt, H.J. van der, J.A.A. Swart and J. Keulartz (2006). 'Nature and landscape planning: Exploring the dynamics of valuation, the case of The Netherlands.' *Landscape and Urban Planning*, 79: 218–228.

West Polesie Biosphere Reserve (by Tadeusz J. Chmielewski)

Chapter 10
Creation of a Bottom–Up Nature Conservation Policy in Poland

The Case of the West Polesie Biosphere Reserve

Tadeusz J. Chmielewski and Jaroslaw Krogulec

10.1 Introduction

As a new EU member state, Poland had to designate Special Protection Areas (SPAs) under the Birds Directive, and to present a complete national list of proposed Sites of Community Importance (pSCIs) under the Habitats Directive by 1 May 2004 – the day that Poland and nine other countries joined the EU. The designation process started in 2000. The preliminary list of Natura 2000 areas covered about 20% of the country's area. However, due to the severe critique of the designation process by local authorities, foresters and water management institutions, the final national list, which was prepared for submission to the European Commission was reduced to about half of its original size, namely to 72 SPAs and 184 pSCIs (7.8 and 3.7% of the country's area, respectively). In response to this proposal, a coalition of NGOs prepared a so-called 'Shadow List', which the Polish government will consider in the next stages of the creation of the Natura 2000 network (Tworek et al. 2005).

According to the Natura 2000 barometer, Poland occupies the last place in the EU regarding the percentage of its national area covered by the pSCIs. The members of the National Council of Nature Protection and many other experts are convinced that the pSCIs network should cover 10–11% of Poland and that the SPAs should cover at least 15% of the country.

However, conditions for nature conservation after the accession of Poland to the European Union have not been very favourable. Biodiversity and landscape protection are considered of minor importance compared to economic development, new investment and the improvement of employment opportunities. Economic growth has gone hand in hand with environmental degradation. The rapid and chaotic

Tadeusz J. Chmielewski

Department of Landscape Ecology and Nature Conservation, Lublin; Agriculture University, Poland

E-mail: tadeusz.chmielewski@ar.lublin.pl; t.chmielewski@pollub.pl

urbanization process in Poland is encroaching upon farmland and natural areas. Forest resource exploitation, in particular, is a very critical factor affecting Poland's biodiversity. Logging increased by 26.7% from 22 hm^3 in 1996 to 30 hm^3 in 2004 (Grzesiak and Domanska 2005). Additionally, very strong pressure developed following major flooding in 1997, to regulate rivers and streams, and this increased wetland degradation. Needles to say, the creation of Natura 2000 met with resistance mainly in the river valleys, many of which still have a natural or semi-natural character.

Moreover, because of their top–down character many nature conservation programmes turned out to be ineffective. Recent bottom–up initiatives of nature and landscape protection by individual regions and individual communities seem to be more promising and productive. This chapter focuses on one of the best examples: the UNESCO Biosphere Reserve West Polesie, created in 2002 and situated in the poorest region, not only of Poland but also of the whole EU (Chmielewski 2005b).

10.2 The West Polesie Biosphere Reserve: Ecology as a Starting Point!

By the end of 2005, 408 nature reserves all over the world had gained UNESCO Biosphere Reserve status, including eight sites in Poland. The youngest and second largest reserve in Poland is the West Polesie Biosphere Reserve, which was created in April 2002 on 139,917 ha. in the most westerly part of Polesie.[1] The central part of the reserve is occupied by the Polesie National Park, while the reserve's western and eastern parts are taken up by the Landscape Park 'Leczna Lakeland' and the Sobibór Landscape Park, respectively.

The most important reasons for creating the Biosphere Reserve West Polesie are the reserve's unique variety of nature, its location on the border of contrasting physiographical and ecological structures, its abundance of lakes, bogs, moors and forests, the presence of many relict and rare species in the reserve, and the unique importance of this region to the European ecological structure.

The initiative to establish a UNESCO Biosphere Reserve in Western Polesie arose in 1991, during an international meeting of scientists and conservation organizations on Polish–Ukrainian cooperation in the field of nature conservation in Polesie. Finally, in 1996, four provincial nature conservationists prepared the first concrete project of spatial organization and Biosphere Reserve border delimitation. This idea met with a high level of approval from local NGOs, which started promoting the project within the community. Members of the Ecological Club UNESCO from the small town of Piaski, near Lublin, were especially active in promotion of this. They were soon joined by members of the Polesie Society, which is an

---

[1] In 2002, an analogous reserve, 'Shatsk', was created in the Ukrainian part of the Polesie Region, and in 2004, a third similar reserve, 'Pribuzskoe Polesie', was established in the Belarusian part of this region (Chmielewski 2005b).

NGO established for the protection of the natural values of the Biosphere Reserve, the sustainable development of the region, and for cooperation with inhabitants of Ukrainian Polesie. From the very beginning, the Ecological Club and the Polesie Society closely cooperated with local governments in preparing popular science seminars, ecological workshops, excursions, and youth competitions for the local inhabitants. Ukrainian guests took part in many of these events. Later, the NGOs prepared similar meetings on the other side of the Bug border river. In addition to these events, the new local journal 'Wonderland Polesie', which popularized this region and its new initiatives, began to be issued.

10.3 Community Involvement

Creation of the Polesie National Park took place at the same time as the formation of local governments in Poland. May 27, 1990 is considered to be the date of the formation of the local governments, because on that date the first municipal elections took place. One of the seven most important tasks included in the Local Government Act of March 1990 concerns environmental and natural protection and water economy.

When the Polesie National Park was created, a majority of the communities' inhabitants, mainly farmers, expressed their disapproval and manifested their discontent about the national park. Misapprehension of concepts and ignorance of legal regulations were not favourable to a friendly adoption of new protection structures, such as a UNESCO Biosphere Reserve, in this region.

The young municipal government, which was equipped with broad competences and rights, and which was partly financially independent, understood very quickly what benefits might be gained from the existence of a National Park and a UNESCO Biosphere Reserve in the territory or in the vicinity of their community. That is why discussions with the inhabitants were initiated concerning the function of the Park, its tasks, competency and relationship with the existing economic, agricultural and social reality. It was stressed that the Park is a golden opportunity for the national, and even worldwide, promotion of natural, scientific, didactic and tourist values of the region; creating opportunities for gaining funds for environmental protection and for the improvement of the living conditions of the inhabitants. These meetings, supported by the friendliness, help and understanding of the Park's staff, especially its board, significantly attenuated the initial distrustful impression.

During meetings of community administrators with the director of the Polesie National Park, the Park's opponents were won over to the idea of nature conservation by the benefits that could be gained from the existence of an area of such highly ranked protection in their territory. The local government understood that by operating according to the statutory activities of a national park, the community might more efficiently perform many of its own tasks as defined in the Local Government Act. In June 1991, these meetings and discussions resulted in an agreement concerning environmental protection in eight communities from two provinces in the region of the Polesie National Park. This agreement put forward the following goals:

- protection of waters against pollution.
- rationalisation of water economy in the economic sense.
- limitation of dust and pollutant emissions into the atmosphere.
- soil protection.
- improvement of public utilities.
- conservation of green areas.
- increase in the region's woodlands.

The priority tasks were

- improvement of water conditions.
- improvement of the sanitary state of communities, including building a waste water treatment plant.
- building water supply systems with a collective water intake (Kopieniak 2005).

Soon each community requested subsidies from various institutions, including the National and Provincial Environmental Protection Fund, the Minister of the Environment, the Eco-fund, the Chiefs of Provinces and other organisations and authorities. The first organised appeals of this type were received warm-heartedly and with understanding by the then authorities who, in turn, provided significant financial means. Communities began building water intakes and water supply systems. These intakes and systems were realised at the beginning in the community towns and villages, and later in smaller village settlements and in groups of recreational built-up areas.

While resolving and restructuring the water supply, there was the problem of an increasing amount of waste water. Because of the funds that the Polesie National Park had at its disposal, new waste water treatment plants were built in Ludwin, Puchaczow, Sosnowica, while the already existing ones in Stary Brus, Hansk, Urszulin, Wierzbica and Cycow were modernised. Collective sewerage systems were also built. An important supporter in this building and renovation work was the Agency of Treasury Agricultural Property, which participated in the costs of dealing with the water supply and sewage disposal problems in the former centres of socialist Collective Farms. In all of these activities, communities could always depend on the favour of the Polesie National Park authorities, which frequently supported these tasks, treating their execution as fulfilling their own statutory duties. Now, all signatories of the Agreement have a sewerage network on their premises.

The example of a solution to the waste water problem and the strong pressure caused by the existence of the Park was followed by further investments, such as individual, domestic, sewage treatment plants. These were built in many households in Hansk, Stary Brus, Ludwin and Sosnowica communities. This solution became very popular in the areas of scattered and settlement housing. It has been estimated that during the next few years a few hundred to even a thousand such mini-treatment plants will be built (Kopieniak 2005).

Natural values and the promotion of the region, following the creation of the Polesie National Park, have stimulated investment towards agro-tourism, especially in Urszulin, Sosnowica and Stary Brus. Copying the tourist and educational activities of the Polesie National Park as their partner, communities began to use the natural values of the region, building tourist and cycling tracks as well as places of shelter and parking areas. Concentrating on natural and agro-tourism may be a chance for the region to develop and prevent economic marginalization of the area. It also creates many new jobs, which is a matter of great importance for this historically poor region.

Taking into account environmental protection as well as the economy, communities began to close down small coal-fired boiler plants, replacing them with modern ones or changing them into a system of furnace oil or propane-butane gas heating. Gas heating will be a prevalent solution, as soon as the region will have a gas pipeline at its disposal. At the moment, communities are talking to Gas Plants, supported by a ready concept of gasification of the whole Wlodawa district and Cycow and Wierzbica communities from the Chelm district. Supplying network gas could enable better air protection, which is another precious value of the region.

A problem that caused a great deal of trouble for years was waste management. This was particularly intense at the beginning of the 1990s when the amount of used packaging increased. Currently, communities are trying to deal with waste disposal by organising village waste collection banks and their disposal into properly prepared and controlled landfill sites. A considerable number of KP-7 type containers have been purchased, and in some communities (Urszulin, Hansk) waste separation has already begun. The target solution of the problem will be the introduction of a common Waste Treatment Plant for the whole Wlodawa district and a common, selective waste collection. This solution will contribute to a radical decrease of illegal waste dumping, which is particularly troublesome for forest areas and post-peat bog wasteland.

It should be stressed that communities have also supported Polesie National Park activities, appealing to the authorities for enlarging the Park, writing positive opinions about the Park's nature conservation plan or regional spatial management and environment protection plan. Cooperation between communities and Polesie National Park has always been characterised by mutual respect, kindness and understanding. However, the communities' achievements concerning environmental protection would not be as effective without a constant, friendly and fruitful participation of the National Park in every undertaking important for the regional inhabitants' quality of life.

In 2000, scientists and NGO volunteers worked out 'The Biosphere Reserve West Polesie Decalogue', which was unanimously passed in the township of Wlodawa during a ceremonial conference of all parties (Chmielewski 2000). The Biosphere Reserve West Polesie Decalogue seeks to:

1. Protect what is valuable and particularly characteristic for the region.
2. Enrich what is impoverished.
3. Renaturalize what is ecologically degraded, and restore what is culturally and morally neglected.
4. Minimize the conflicts.

5. Solve the problems in harmony.
6. Counteract the threats.
7. Use the potentials in a better way.
8. Support and develop local initiatives.
9. Look for external financial support for supralocal tasks.
10. Promote the values and harmonious development of the region at the country, European and world level.

Local communities started to feel as though they are integral partners of the project and considered it a chance for trans-boundary cooperation and international promotion of the natural, health and recreational values of the region. Biosphere Reserve targets have proved to be in agreement with local community expectations. The need to establish the Biosphere Reserve was included in the Western Polesie communities' development strategy.

10.4 Protected Areas System Management in the West Polesie Biosphere Reserve

As indicated above, the work on sustainable management of the natural resources in the region started within the Lublin scientific community, and the results of this work were presented to local government authorities. It was agreed that a nature resources management system in the Biosphere Reserve should be based on 4 interconnected subsystems (Chmielewski and Domagala 2005):

- **Diagnostic Subsystem**: this subsystem focuses on acquiring knowledge of the spatial conditions and processes, ecosystems and landscape composition of the region.
- **Planning subsystem**: on the basis of the information provided through the diagnostic subsystem protection plans for the national park and the landscape parks, sustainable development strategies for the communities of West Polesie are worked out.
- **Decision subsystem:** on the basis of the protection plans and local spatial management plans, administrative decisions concerning protection actions for specified types of ecosystems are being issued. These can relate, for example, to water damming, stand reconstruction, plant succession control on moorland, as well as decisions concerning architectural objects, tourist facilities, the creation of tourist routes and educational paths.
- **Control subsystem:** this subsystem is a logical follow-up to the previously detailed subsystems, in which several aspects including building inspection, environmental quality control and nature monitoring are combined to monitor and evaluate the management in the West Polesie Biosphere Reserve.

10.4.1 Diagnostic Subsystem

The key actors in this subsystem include universities and research institutes, the Scientific Councils of Polesie National Park, Chelm Landscape Parks and Lublin Landscape Parks.

The cooperation has been developing very well, especially in case of the Polesie National Park. In the area of the Biosphere Reserve, numerous national and international research programmes are being realised. Each year, 27–38 research programmes are realised in the Polesie National Park alone, and in the Polesie National Park's 15-year existence, research carried out on its territory has generated 9 PhD theses and over 300 MSc theses from 130 scientific units in the country and abroad (Piasecki 2005). As a result, the area of the Biosphere Reserve West Polesie is one of the better-studied ecological landscape systems in Poland (Chmielewski 2000; Radwan 2002). Good scientific recognition of ecological processes occurring in the area under protection is the basis for correct management of resources and natural space values of the region.

10.4.2 Planning Subsystem

The following actors play a decisive role in the functioning of the planning subsystem:

- The Director of the National Park– concerning the area of the national park,
- The Chief of the Province (Voivode) – concerning landscape parks and nature reserves.
- Individual local governments.

Coordination of planning work dealt with by the above individual units is realised by a system of mutual agreements and opinions. Protection plans have been worked out by the governmental nature protection service; however, local government representatives also took part in the process of the plans' preparation. Coordination is not an easy task, and divergent intentions and conflicts of interest often occur. A joint approach in establishing the 'Strategy of Nature Protection and Sustainable Development of the Biosphere Reserve' would be a good solution to the problem for all parties. Up until now, the communities of the West Polesie have drawn up their strategies of sustainable development by themselves. These strategies are not fully compatible with the plans of the national and landscape parks' nature protection, or with the requirements of the Natura 2000 sites protection in the region.

In accordance with the criteria on the area of the West Polesie Biosphere Reserve specified in the Habitat Directive and the Birds Directive, 6 habitat sites covering a total area of 92792 ha. and 9 bird sites with a total area of about 45000 ha. should be created in the biosphere, but most of the area of the sites will cover the already existing forms of nature protection. Interactions between the competencies concerning the management of the Natura 2000 sites and other existing protected areas of the region, as well as the plans for the special development of the communities, are not legally settled. This has caused several problems for regional and local authorities of nature protection. Solving this problem is not, however, the duty of the communal or provincial authorities, but the task of the European Commission and the government of the Republic of Poland.

In the present situation, scientists and authorities of the West Polesie Biosphere Reserve are trying to work out the best methods of nature conservation planning, in agreement with the sustainable development of the region. These methods are based on the latest achievements of landscape ecology and on the results of interdisciplinary diagnostic studies (Chmielewski 2001a; 2005b).

10.4.3 Decision Subsystem

The main role in the functioning of the decision subsystem is played by local governments and concerns the development of the area, the use of the natural resources, and protection and planning within the green areas. Decisions concerning the economy of state forests are issued by forest managers, and in private forests by local government. However, the main authority concerning the biodiversity protection belongs to the Chief of the Province (Voivode), and in the territory of a national park, to the park's Director.

In 1992, the West Polesie Provincial Nature Conservator, as a representative of the Lublin Chief of the Province (Voivode), made the decision to commence a pioneer wetland restoration project. Two years later, the Director of the Polesie National Park did the same thing. The positive results from the realisation of these projects became an inspiration and a model for analogous actions in other regions of the country (Chmielewski 2001b; Chmielewski 2003).

In the decision-making process, and in investment realisation, the harmonious cooperation of the national park and the landscape parks nature protection services with local governments is important. The beginnings of the cooperation were difficult in West Polesie because people were afraid of economic bans and restrictions connected with the introduction of various forms of nature protection, especially with the creation of a national park. Reality, however, showed that the national park greatly promoted the region and a significant improvement of the inhabitants' living conditions. Many key investments for the region were realised by communities working together with the national park. In the case of Biosphere Reserve West Polesie, the realization of several wetlands restoration projects has been particularly important, including the Piwonia River and the Nadrybie- Bikcze-Ciesacin Lake Complex Restoration, which has a total area of 600 ha. This was a pioneer project in Poland, realized between 1992 and 1995 (Chmielewski 2003). Functioning of the decision subsystem has not been satisfactory in this region so far, because many decisions, particularly concerning the new economical investments, neglect the ecological conditions or marginalize them.

10.4.4 Control Subsystem

Of all the subsystems, the control subsystem is the most problematic, especially the control of the realisation of the spatial development plans and the monitoring of environmental transformation, although this situation is gradually improving. In particular, a network of biodiversity and landscape diversity monitoring, as well as urban monitoring, is urgently needed. It is also necessary to systemize the collection of the monitoring data by a common introduction of Spatial Information Systems based on advanced computing technologies. Such systems are one of the key instruments in facilitating effective protection and sustainable management of resources of unique natural value areas on a world scale. However, these systems are only beginning to be created in the richest regions of Poland. In the meantime, Biosphere Reserve West Polesie is situated in the poorest region of not only Poland but also the whole European Union. Technical equipment and training of the personnel for the creation of the West Polesie Biosphere Reserve Spatial Information System is one of the tasks that should be organised within the confines of the support programmes for the poorest region of European Union.

10.5 Conclusion

The bottom–up initiatives of nature and landscape protection in the west of Poland can be considered successful and productive. The initial resistance by the local communities was met with a great deal of understanding by the scientific and ecological initiators of the West Polesie Park and the intended encompassing UNESCO Biosphere Reserve. Emphasis was placed on the role the natural values of the area could play in the social and economic development of the area, in addition to the preservation of the ecological and spatial characteristics of the area. Opponents were ultimately won over to the idea of nature conservation by the benefits that could be gained from the existence of such an area.

The promotion of the region and its natural values has stimulated investment operations towards agro-tourism. Concentrating on natural and agro-tourism offers an opportunity to the region to develop and to prevent economic marginalization of the area, which is of great importance for this historically poor region. Further initiatives were undertaken to deal with the problems of waste water, waste management and energy supply that benefited both the community and the environment.

Cooperation between the communities and the Polesie National Park has always been characterized by mutual respect, kindness and understanding. Although, the communities' achievements concerning environmental protection should be considered effective, the decision and control subsystem still needs improvement. In the former, new economic investments neglect the ecological conditions or marginalize them, and in the latter, a network of biodiversity and landscape diversity monitoring as well as urban monitoring is urgently needed.

To effect these improvements, all activities concerning nature conservation policy in the West Polesie Biosphere Reserve should be focused on:

1) The common use of landscape ecology methods and techniques in the diagnosis of changes in ecosystems and their landscape complexes, and in the evaluation of interrelations between biological diversity and the ecological landscape structure.

2) The improvement of nature protection planning methods by means of:

   - evaluation of ecological effects of previous protection plans;
   - organization of workshops and seminars with the participation of local authorities, focusing on the establishment of administrative decisions based upon deep understanding of and respect for the rules of the functioning of nature;
   - development and support of NGO activities, which aim at the promotion of human sustainable development and life-style in harmony with pristine nature.

3) The organization of a smoothly functioning system of nature and urban monitoring, and development of social control methods to maintain standards of environmental quality.

References

Chmielewski T. J., ed. (2000). The West Polesie Transboundary Biosphere Reserve – nature and culture harmonisation project. Polesie National Park, Lublin Province, Lublin – Urszulin: 1–120 (in Polish, English summary).

Chmielewski T. J. (2001a). Spatial Planning System Harmonising Ecology and Economy. Lublin University of Technology, vol. 1–2: 1–294 and 1–146.

Chmielewski T. J. (2001b). Scale and directions of landscape transformation on the Leczna-Wlodawa Lakeland: from degradation to the restoration [In: Roo-Zielinska E., Solon J. (eds.): Between Geography and Biology: Environment Transformation Research]. Institute of Geography and Spatial Management. Polish Academy of Science. Geographical Studies, vol. 179. Warsaw: 103–116 (in Polish, English Summary).

Chmielewski T. J., and J. Krogulec (2003). Ten years of experience in implementation of environmental engineering in the protection of biodiversity. The case of Lublin region (CE Poland). In L. Pawlowski, M.R. Dudzinska and A. Pawlowski, eds. Environmental Engineering Studies: Polish Research on the Way to the EU. New York, Boston, Dordrecht, London, Moscow: Kluwer Academic/Plenum Publishers, pp. 431–442 (in English, Polish summary).

Chmielewski T. J. (2005a). Nature conservation plans as an element of ecological space management system in the hydrogenic landscapes [In: Chmielewski T. J. (ed.) 15 years of the Polesie National Park]. Ministry of Environment, Polesie National Park. Warsaw – Lublin: 115–126 (in Polish, English summary).

Chmielewski T. J., ed. (2005b). The West Polesie Biosphere Reserve: values, function, perspective for sustainable development. Polesie National Park, Lublin Voivode. Lublin – Urszulin: 1–168 (in Polish, English summary).

Chmielewski T. J., and R. Domagala (2005). Integrated system of ecological space management: the case of the West Polesie Biosphere Reserve. Landscape Architecture; Wroclaw; 3–4: 60–66 (in Polish, English summary).

Grzesiak M., and W. Domanska, eds. (2005). Environment 2002. Annual Report of Central Statistical Office in Poland. Warsaw: 1–540 (in Polish).

Kopieniak M. (2005). Activity of local authorities of Polesie National Park for building harmony between nature and economy. [In T.J. Chmielewski, ed. 15 years of the Polesie National Park]. Ministry of Environment, Polesie National Park. Warsaw – Lublin (in Polish, English summary).

Piasecki D. (2005). The Polesie National Park: history of establishing and 15 years of activity [In: Chmielewski T. J. ed. 15 years of the Polesie National Park]. Ministry of Environment, Polesie National Park. Warsaw – Lublin (in Polish, English summary).

Radwan S., ed. (2002). The Polesie National Park. Ecological Monograph. Morpol, Lublin: 1–272 (in Polish, English summary).

Tworek S., M. Makomaska-Juchiewicz, J. Perzanowska and G. Cierlik (2005). Evaluation of Nature 2000 network proposal in Poland. Report. Institute for Nature Conservation Polish Academy of Science, Kraków: 1–21 (in Polish).

Chapter 11
Nature Conservation in Poland and the Netherlands

Bottom–Up or Bottom Line?

Henny van der Windt

11.1 Introduction: West Polesie in Context

Poland and many other new EU members are interesting because their traditions with respect to agriculture, nature conservation and policymaking differ from most of the older EU members. At the same time there are several similarities, with examples being the use by national governments of a top–down approach linked to EU directives, and the attempt to overcome implementation problems by working with stakeholders, local governments and others.

One of the remarkable points in the paper about West Polesie, is its description of the conservation organizations' efforts to found a relatively large and diverse Biosphere nature reserve using a bottom–up approach. Biosphere reserves are not covered by an international convention. They are defined as ecosystems promoting solutions to reconcile the conservation of biodiversity with its sustainable use. As such, they have a conservation function (landscapes, ecosystems and species), a development function (sustainable economic development) and a logistical function (research and education). This particular case involved conservationists' demands and restoration goals regarding ecological structure and processes, as well as aims and goals regarding water management and economic development in the region, especially with regard to recreation. This approach was also a reaction to the limited success of the plans of the central government to establish a network of nature reserves. This new approach involved local stakeholders, and the authorities looked for new procedures. Thus far, the cooperation between these stakeholders and authorities has improved, but there are still problems to resolve. According to the authors, the main problem is maintaining the quality of the area in terms of biodiversity. They suggest setting up a new planning and management system for the area.

Henny van der Windt
Science & Society Group (SSG), Faculty of Mathematics and Natural Sciences,
University of Groningen, Netherlands
E-mail: h.j.van.der.windt@rug.nl

J. Keulartz and G. Leistra (eds.), *Legitimacy in European Nature Conservation Policy:*
Case Studies in Multilevel Governance, 149–157. © Springer 2008

To be able to characterize and evaluate the bottom–up decision-making process concerning the Polish nature reserve and its legitimacy, some kind of theoretical scheme is needed. Such a preliminary interpretative scheme will be derived from recent literature (Susskind and Cruikshank 1987; Woltjer 2000; Scharpf 1999; Van Tatenhove et al. 2000). Furthermore, I will present comparative findings from a Dutch case study in order to place the Polish case in perspective.

11.2 Types of Decision-making and Legitimacy

Because the term 'bottom–up' is central to this paper, I will firstly define it before discussing other criteria and aspects of participative-deliberative decision-making. Finally, I will raise some additional issues concerning decision-making, such as the efficiency and quality of the outcomes of the process.

11.2.1 Bottom–Up

The issue of participation is a far from recent development. As early as the 1960s, many analysts in Europe and the USA were looking for ways to enrich parliamentary democracy by employing a more bottom–up method of decision-making. In her famous 'ladder of citizen participation', Arnstein (1969) classified several possible relationships between the state and citizens' organizations (see Table 11.1). On the first three rungs of this ladder, which Arnstein labelled manipulation, therapy and information, there is very little participation by citizens or citizens' groups. On the following three rungs, consultation, placation and cooperation, participation is found to be symbolic. On the final two rungs, delegation and self-governance, citizens share power with the governmental institutions.

Table 11.1 Ladder of participation (Arnstein 1969; see also Voogd 1995; Woltjer 2000), related to the role of regional or local actors and the role of government

Form and extent of participation	Role of regional/local actor	Role of government
1. Self-governance	Initiative, making choices	Supporting
2. Delegation	Taking decisions within frameworks	Offering opportunities within frameworks
3. Cooperation	Making choices together	Making choices together
4. Placation	Discussing, offering ideas	Offering possibilities for discussion
5. Consultation	Giving opinions	Offering solutions and asking for replies
6. Information	Consuming information	Offering information
7. Therapy	Doing almost nothing but ready to undergo therapy	Doing almost everything
8. Manipulation	Doing almost nothing but ready to undergo manipulation	Doing almost everything

11.2.2 Interactive Decision-making

Bottom–up decision-making has consequences for both the role of the societal actor and the government, as is shown in Table 11.1. Compared to the more usual top–down approach, this form of interactive policymaking also requires changes in the use of policy instruments (less legal enforcement, more discussion and negotiation), another style (less formal, more flexible, more transparent) and sometimes another, extended, problem definition (including visions and values, other types of knowledge).

Thus, although the positions and relationships of the different actors are the main aspects related to interactive decision-making, and consequently to procedural legitimacy, there are at least three other issues at stake (see Table 11.2). Firstly, the resources that actors choose to use, varying from legislation to financial means or trust, can make a significant difference. Furthermore, the style of government and the character of the procedure are important, for example, in relation to transparency, fairness and openness. Finally, we have to consider the aims and problems that were accepted as legitimate during the different stages of the decision-making process, and how they were incorporated into the process. Society and governments often have to deal with several problems at a time, or with several dimensions of a problem. Integral solutions within new modes of governance are needed, and these are usually found using the problem-solving capacity already existing within the society. For example, issues of sustainability or landscape management have to deal with social, scientific, aesthetic, ethical and economic considerations. Here, facts and visions count, but their relevance may differ during the decision-making process. In general, decisions are seen as having a high degree of procedural legitimacy if:

Table 11.2 Three modes of decision-making and their characteristics

Aspect	Top–down	Corporative	Deliberative-participative
Actors and influence	Central government is dominant	Main stakeholders have much influence	Many actors; influence not determined by power
Style of government	Forcing and convincing; not transparent, formal, not flexible	Negotiations within existing networks; transparent for main actors, flexible and not merely formal	Deliberations, including visions and values, learning, transparent, flexible, not merely formal
Means and resources	Legislation, existing policy plans, scientific knowledge	Finances, goods, people (members), interests	Communication, arguments, trust
Aims and issues	Product orientation, fixed standards and issues	Product and process orientation, fixed issues	Process orientation dominant, wide range of issues, including values

- all relevant actors have a considerable influence.
- the government is willing to adapt its original decisions.
- the government stimulates openness, trust and fairness.
- a wide range of problems, including long-term issues, are accepted.

Under the umbrella of interactive decision-making, at least two modes can be distinguished – corporative and deliberative. In the corporative mode, the number of stakeholders and the openness and transparency of the process are limited, while negotiations, financial means, interests and the outcome in terms of a product rather than a process are more important as compared to the deliberative mode (Table 11.2). However, even deliberative modes of decision-making can be different. For example, the power and influence of participants may vary considerably.

Finally, decision-making is not just a matter of participation and process, but also has certain aims with respect to the product. The outcome of the decision-making process should be acceptable in terms of quality, efficiency and effectiveness.

11.3 A Dutch Case Study: The Green River Plan

Just as in many other European countries, the Dutch government decided to create an ecological network with core areas and corridors. The provincial authorities were charged with its implementation. In many places this led to resistance from farmers and other citizens. Because of developments in national water management in the 1990s, related to a growing awareness of the effects of climate change, most attention was paid to wet corridors. Both at the national and the regional level, conservation organizations took the initiative, presenting plans to restore old river and brook systems in order to stimulate the migration of organisms, but also to achieve aims such as flood safety and environmental quality. At the regional level, the demands of fisheries, the quality of the landscape and recreation were also taken into account, for example, in the so-called Green River plan for the northern part of the Netherlands (Van der Windt et al. 2006), which enabled water flow from the high sandy plateaus to the city of Groningen and thence to the sea.

The Green River plan won the support of the Royal Dutch Tourist Organization ANWB, the largest tourist organization in the Netherlands, and similar ideas arose within governmental organizations. However, there was also resistance and scepticism. Farmers were afraid that they would be forced to move or to change their management style, and industrial parties feared that these proposals would block or delay the plans for an industrial zone near the city of Groningen. The regional water authority responsible for the area was sceptical of the idea of water management through the restoration and construction of wet ecological corridors. Moreover, despite the interest and enthusiasm of some local authorities, other provincial and communal politicians were not as keen because of the costs and expected negative effects on agriculture and industrial planning. The province of Groningen was fully aware of the different aims, visions and interests related to the Green River plan and similar initiatives. To ensure commitment and create a consensus, the province initiated a number of workshops for input from the other authorities and stakeholders, assisted by several consultants.

Gradually, the Green River plan became an important issue, linked with other concerns within a rather complicated process of decision-making. It was not only a multi-issue, multi-sector and multi-actor process, but also involved many levels of government. Community, provincial, regional and state governments became involved, bringing different visions for the region and its watercourses, and different notions of organizational responsibility. The stakeholders involved also displayed different interests, perceptions and values concerning nature and water. As expected, conservationists preferred wilderness, aiming to keep the environment as undisturbed as possible. The authorities of the city of Groningen and the water authority preferred a functional approach to nature, that is, a natural environment which also supported human interests such as recreation and transport. The preferences of farmers and provincial authorities were somewhat mixed, incorporating both the functional and wilderness perspectives, but also including an arcadian view of nature which focused on traditional agricultural landscapes possessing a high degree of biodiversity. Workshops and other communicative efforts on the part of the authorities contributed to a bridging of the contrasting viewpoints, especially those of the nature conservationists and water authorities. The water authority required appropriate places for water storage in order to prevent the possible flooding of urban areas, and it became aware that this storage requirement could be achieved through the restoration of brook valleys. The conservation organizations accepted that these restored brook valleys could have a storage function which would prevent flooding. The farmers' preferences depended on the precise area in question. In some areas of the brook valley system, farmers were willing to sell agricultural land to nature management organizations, or to manage ecological zones along the brooks in a nature-friendly way in exchange for payment. In other areas they asked for assurances that they could continue to develop rationalized forms of agriculture.

With hindsight, the role of the province was somewhat ambiguous, despite its attempt to reach a wide consensus. The province itself had some reservations because the Green River plan was not completely in line with its nature conservation and spatial policy. These policy plans did not leave much room for the development of riparian systems along the central axis of the proposed brook system. As a compromise, the province proposed establishing a small ecological corridor just, 50 metres wide, instead of the 250 metre-wide zone advocated in the Green River plan. Meanwhile, the city of Groningen, which had the main responsibility for the industrial zone, was ambivalent. After the intervention of the Green Party in the city council, the city was forced to accept the Green River plan in principle, with an ecological zone width of 150 metres. It is expected that this will be a sustainable compromise.

11.4 Evaluation

11.4.1 Actors and Influence

In the Dutch case, several layers or networks of decision-making can be distinguished, each related to a few central problems or a certain area. The influence of the different actors has been dependent on these layers and networks. Nature conservation NGOs were successful in changing the original plans, and in developing their own plans for new riparian areas and brooks as part of a Green River plan. Together with national and regional authorities, consultants and other stakeholders, they reached a consensus on a restoration plan for a brook system, and it is in this sense of building a consensus that the term bottom–up is appropriate. However, only well-organized groups were engaged in the debate, with individual citizens being left out of the process. Most of the other actors opposed to the nature conservation NGOs, including farmers, industrial lobbyists and water authorities, also succeeded in achieving their main goals while seeking common ground. As a result of the framework established by the regional governments for the discussion of the plans and their implementation, none of the actors involved dominated the process.

In the Polish case, different decision-making layers, each related to particular areas and sometimes to specific responsibilities, can also be seen, such as the communities, the central government, the province and the national park. Just as in the Netherlands, nature conservationists were ultimately successful. Because of different concerns, specifically the aim to establish a biosphere reserve and the existence of the national park, the demands of the nature conservationists coincided far more with at least some sections of the authorities involved. Here again the term bottom–up is applicable, although the bottom level had a strong nature conservation element, and it remains uncertain how far the conservationists' influence and participation will extend. The nature conservationists are not convinced that the quality of the nature reserve will be preserved in the future. Just as in the Netherlands, opposition arose among farmers and in some communities, and probably also came from water authorities and those concerned with forestry. However, compared to the Netherlands, citizens and scientists played a more active role in the West Polesie decision-making process.

11.4.2 Styles of Government

In the case of the Dutch Green River plan, various levels of government created room for negotiations and discussions within the networks, and organized informal workshops for the main participants. Local initiatives such as the Green River plan were taken seriously by most governments, and attempts were made to combine these demands with aims from other sectors, such as farming and water management. It is notable that independent facilitators were used at several stages of the process.

In Poland, communities and other governmental authorities were also open to negotiation and discussion. It seems, however, that in this case much more attention was paid to education, for example, by holding ecological workshops and conducting excursions. Similarly to the Dutch case, efforts were made to combine the aims of nature conservationists with those of other sectors, such as waste and water management, and recreation.

11.4.3 Means and Resources

In the Netherlands, one of the problems for the nature conservation NGOs was that their plans were not in line with the formal policy agenda. As a result, formal policy and legislation were important sources of power for the governments involved in the negotiations. At the same time, the various governments realized that they were very dependent on several of the groups involved, and so attempted to create a trustworthy atmosphere. Finally, to compensate the farmers for damage, financial means were used to facilitate the decision-making process. Science and scientists did not play a visible role at these levels, although the ecologists provided the concept of an ecological corridor.

Considering the rapid change in the appreciation of the nature reserve plan expressed by some communities, the Polish case seems to be much more informal. The importance of trust and consensus is expressed by 'The Biosphere Reserve West Polesie Decalogue', which stresses the need for harmony, consensus and local initiatives. However, ecological knowledge, ecological advice and ecologists seem to be much more important resources than in the Dutch case.

11.4.4 Aims and Issues

As mentioned above, in the Dutch case some policy plans altered, although some government levels did not show a great deal of political flexibility. Also, we saw that the plans for water management and nature conservation were integrated into a multi-sectoral and multi-issue approach, and the themes of industry, town planning, the fisheries, landscape and nature protection, and climate change were all taken into account. Furthermore, attention was paid to economic, ethical and aesthetic issues. Finally, we attempted to look at the different stages of the decision-making process, and at different scales in space and time.

In the Polish case, this shift to multi-sectoral and multi-dimensional approaches can also be recognized. For example, in 'The Biosphere Reserve West Polesie Decalogue', cultural and moral values are mentioned, and a harmonic development of the region is proposed which includes some room for sustainable economic growth. Furthermore, the stated aim is to recognize the economic, cultural, social and natural potential, and to examine the relationship between the situation inside and outside the reserve. Good examples of multi-sectoral approaches are the combinations of tourism and nature conservation with the simultaneous improvement of the quality of water, biodiversity and the landscape.

The paper also mentions a plan to make a clear distinction between 'real' nature management, which is related to ecology, and other types of management. This entails a division into four spheres:

1. Diagnosis: based in ecology and landscape science, and carried out by scientists.
2. Planning: carried out by nature conservation organizations, together with communities, scientists and planning experts.
3. Decision-making: concerning certain ecosystem protection and management/use of the area. Here the responsibilities are unclear.
4. Control: monitoring and inspection. Here, too, it is not clear who is responsible.

Thus, both in space and time, different parties are responsible for the management of the area.

11.4.5 Other Aspects: The Bottom Line

In the Dutch Green River case, the outcome of the process is regarded as being better than, or equal to, the original plans with respect to nature conservation, urban-industrial quality, landscape quality, recreation, fisheries, agriculture and water management; with most participants being satisfied. As such, the quality of the product is sufficient and the interactive process involved in its creation has been shown to be effective. It is much more difficult to judge the efficiency than the effectiveness of the process and the quality of its outcomes, but what is certain is that it took a great deal of time and energy to reach this consensus.

The outcome of the Polish West Polesie case is even harder to judge. With respect to nature conservation, water quality and recreation, the plan seems to have brought an improvement; with the main players in nature conservation having established a scientifically based bottom line through which to assess ecological quality. The results with respect to such areas as agriculture or forestry are not known, and although all the participants seem to be satisfied, any comments regarding the efficiency of the process would only be preliminarily.

11.5 Concluding Remarks: West Polesie in Context

The introduction formulated four criteria for a high degree of procedural legitimacy: influence of all relevant actors, willingness of government to adapt decisions, government's willingness to stimulate openness, trust and fairness, and the possibility for deliberation on a wide range of issues. As shown above, in the Dutch Green River case many but not all of those affected were included in the discussions. Government levels more or less showed a willingness to change plans, while some government levels stimulated openness and trust – within networks and also using other means such as money and legislation – and there was room to discuss a wide range of issues; resulting in a multi-sectoral decision-making process. This process should be labelled as a mixture of top–down, corporative and deliberative-participative decision-making.

How then should we describe the process in Poland? Obviously, the general situation is different from that in the Netherlands. Firstly, Poland does not have the tradition of citizen participation or the physical planning experience of the Dutch in the 1960s. Furthermore, Polish farmers did not produce for the world market for a long time, and retained more traditional methods. Finally, forestry is also much more important in Poland. Unfortunately, not all the information regarding the decision-making process is provided in the paper. Nevertheless, I will attempt to label the process as in the Dutch case.

To start with, as far as can be derived from the paper, many actors were involved, as well as citizens, but probably not all stakeholders. Several levels of government were also willing to adapt the new plans as developed by nature conservationists, and trust and openness were important issues. It is unclear to what extent national or EU legislation played an important role in the background. It appears that different issues were discussed and integrated, and even values were at stake. In contrast to the Dutch case, no independent facilitators were used. Thus, in conclusion, the process in Poland can be seen as a mixture of top–down and deliberative decision-making. Last but not least, because of the relatively large role of science, scientists and other experts in the founding of the West Polesie, not only procedural but also substantive legitimacy sources were considered important.

References

Arnstein, S. (1969). 'A ladder of citizen participation.' *Journal of the American Institute of Planners*, 216–223.

Scharpf, F. (1999). *Governing in Europe*. Oxford, Oxford University Press.

Susskind, L., and J. Cruikshank (1987). *Breaking the Impasse*. New York, Basic Books.

Tatenhove, J. van, B. Arts and P. Leroy, P. (2000). *Political Modernization of the Environment*. Dordrecht, Kluwer.

Voogd, H. (1995). *Methodologie van ruimtelijke planning*. Bussum, Coutinho.

Windt, H.J. van der, J.A.A. Swart and J. Keulartz, (2006). *Natuurbeleid, schaakspel op verschillende borden*. Haren/Wageningen/Den Haag, RU Groningen/WUR/NWO.

Woltjer, J. (2000). *Consensus Planning*. Aldershot, Ashgate.

The North York Moors (by Michael Carrithers)

Chapter 12
Conservation in Context: A View from Below

Implementation of Conservation Policies on the North York Moors

Elizabeth Oughton and Jane Wheelock

12.1 Introduction

This chapter looks at the perceptions that a small group of stakeholders have of the conservation activities in the National Park in which they live. It is a bottom up study which explicitly looks at the relationship between the social and natural environments.[1] It is not a study of deliberative democracy and the policy making process. It is a study of how policies are experienced by those on the ground; how policy is legitimated, or not, through practice. What we are looking at is the total effect of all relevant policies and their outcomes; it is difficult if not impossible to tease out just one policy without reference to the rest. The study contributes to the debate on policy legitimacy because it explores the real world of policy implementation and policy interactions as these affect people who live and work in an area of nature conservation.

We examine views of legitimacy by exploring people's claims (entitlements) on the landscape. First, we look at the ways in which policy implementation affects such claims and the ability of local people to achieve their objectives with respect to the goods and services that the environment offers. Second, we explore the ways in which actions by locals can support or undermine the conservation objectives of international and national policy making bodies.

We start this chapter with a description of the empirical study on which the analysis is based, a summary of the conceptual basis of our argument and an introduction to the North York Moors.[2] This is followed by a discussion of the way

Elizabeth Oughton
Centre for Rural Economy, Newcastle University
E-mail: E.A.Oughton@newcastle.ac.uk

[1] See Adger et al. 2003 for the necessity of interdisciplinarity across scales to develop effective environmental governance.

[2] The project 'Developing tools for interdisciplinary research' was funded by three British national research councils, ESRC, BBSRC and NERC under the interdisciplinary Rural Economy and Land

in which European conservation directives are implemented in Britain and this in turn is related to regional conservation policies. Section four explores in more detail the knowledges that underpin claims within the regulated landscape. Section five illustrates the argument with reference to a case study of conservation of the heather moorland. In the final section the conclusions and implications for the development of conservation policy implementation are discussed.

12.2 The North York Moors

The focus of the study is a river catchment, the Esk, in the north of England. The River Esk lies on the north eastern edge of the North York Moors, the largest heather moorland in Europe covering approximately 49,900 hectares. It is an area designated as a Special Protection Area (SPA) (meeting obligations under Article 4 of the EU's Birds Directive EC/79/409), a Special Area of Conservation (SAC) (meeting with obligations under Article 6 of the Habitats Directive 92/43/EEC), and Sites of Special Scientific Interest (SSSIs) (UK designation). The area was designated a National Park by the British government in 1952 and is regulated by the North York Moors National Park Authority (NYMNPA) which has responsibilities of both conservation and public enjoyment. Under the 1995 Environment Act these responsibilities were extended to include '...to seek to foster the economic and social wellbeing of local communities.'(NYNMPA 1998:12). In 1998 ninety percent of the moorland area was designated as an SSSI, '...The North York Moors [are] important for a number of species of national and European importance including otter, merlin, golden plover, great crested newt and white-clawed crayfish.' (NYMNPA 1998: 32).

The immediate river valley of the Esk comprises small mixed farms, but the landscape is dominated by the rolling heather moorlands which form the upland catchment. The Esk is of importance for its salmon and trout fishing and contains a population of threatened freshwater pearl mussels. The multiple uses of this moorland landscape provide the basis for livelihoods and local economic development. Economic restructuring at the market economy scale has impacted on livelihoods in ways that indirectly affect the natural environment of the North York Moors. This is particularly due to the adoption of labour saving methods of production. Time and cash pressures on those in employment, business or farming have had a negative effect on maintaining the natural environment. Yet the moorland landscape is essential to tourism and to attracting those of working age and retirees to live in the area. The moorland in the area is managed for grouse shooting and low intensity sheep grazing.

Our approach is essentially one of public ecology (Robertson and Hull 2003). Using established research networks, we initially contacted a wide range of individ-

Use Programme. The work of the project was undertaken with our two colleagues Louise Bracken and Katy Bennett, but they are not responsible for the interpretation here.

ual and institutional stakeholders to recruit participants to four themed meetings and a feedback seminar at the Danby Moors Visitor Centre in the heart of the catchment. Between 9 and 15 people attended each meeting, drawn from a range of institutions including the EA (Environment Agency), NYMNPA (North York Moors National Park Authority), DEFRA (Department of Environment, Food and Rural Affairs) local residents, parish counsellors and a member of the Court Leet.[3] They included professional environmental scientists, local naturalists, and others with experience or training in non-scientific arenas. We asked that people commit themselves to attending the full series of meetings, which may have contributed to the difficulty of recruiting individual farmers, although a couple of landowners did attend, together with a representative of the CBLA (Country Business and Landowners Association). The four themes of the series were: knowledges, entitlements, livelihoods and regulation. Each meeting was run as a focus group led by a nominated facilitator supported by others from the research team. The meetings were recorded and the transcripts used for analysis, supplemented by individual conversations between team members and participants over lunch and tea breaks. The core discussions took place over a full afternoon. The range of stakeholders at each discussion provided us with a microcosm of the process of public negotiation of environmental issues.

Our study was designed to capture systematically the social and environmental processes at work in the development and conservation of the natural resources. The roots of our conceptual thinking derives from Leach et al. (1999) and their work on environmental entitlements, itself taken from Amartya Sen's entitlements analysis (Sen 1999). This has been further developed by Oughton and Birch (2004). Leach et al. posit that individuals make claims on the natural environment, and that the ability to achieve those claims is affected by institutions at various levels. This is what Leach et al. call 'environmental entitlements'. Summarising a complex model: entitlements arise through the ability of actors to draw upon their assets or endowments. The transformation of endowments to entitlements is crucially affected by the institutional context. A wide range of assets have been considered as endowments by different authors. Endowments may include physical, financial, human, political and social assets. In this analysis we focus on those endowments which are made up of values, knowledge and power and explore them with particular reference to the ways in which the implementation of conservation regulations, international, national and regional, are legitimated within a particular local community. A community is not of course a homogeneous entity but composed of individuals who may have very different perspectives and desires in relation to the use of natural resources.[4]

---

[3] The Court Leet has its origin in the medieval period. The Danby Court Leet is one of the few remaining in England and has responsibility for managing the approximately 50% of the common land of the moor. Commoners have grazing rights, the owner of the soil has mineral and shooting rights.

[4] The organic and dynamic nature of environmental entitlements is illustrated more clearly in the discussion below (Section 12.5).

Each stakeholder, be they an individual or a group, will have a range of motivations which they prioritise in line with their values. The institutional context is crucial here, both in terms of value formation and choice (Folbre 1994). The relative power of stakeholders to achieve their entitlements will also depend on this institutional context. Livelihood creation was prioritised by many participants. Other motivations included biodiversity conservation and enjoyment of the natural environment: indeed there was a conviction that all those who live, work or visit the Moors share a love of the North York Moors environment and landscape. Belonging to a thriving local community was also important to many; and people wanted to see transparency in processes of governance.

It is important to distinguish two distinct aspects of the institutional context.[5] First there are institutions in the sense of organisations. The organisations range from international (EU Environment Commission) to local (Parish Council). They may have a focus on environment or economy, or the effect of their activities on the environment may be quite incidental (Olsson et al. 2004 illustrate the very complex nature of these relationships). In our analysis we limit ourselves to the organisations that have a direct effect upon the conservation management of the natural environment. The second meaning of institutions is the more informal 'rules of the game' (North 1990). Markets operate according to one set of rules, while habit and tradition operate according to very different rules. In more everyday parlance, this means 'how things are usually done'. This relates to the knowledges (formal or informal) that different people draw upon. These rules of practice may therefore be more or less transparent.

12.3 Conservation Regulation in England[6]

At the time of writing (October 2005) the Natural Environment and Rural Communities Bill is under consideration by the British Parliament. The Bill is a response to the Rural Delivery Review of October 2003, known as the Haskins report. The objective of the Bill is to provide a more rational structure for the implementation of rural policy affecting people and the environment in rural Britain. Thus the description given here of the implementation of nature conservation deriving from the European Union relates to the situation in the recent past. The Bill will bring about a number of changes in the administrative structures responsible for applying conservation regulations. An integrated agency to be known as Natural England will combine the functions of English Nature (EN) the Rural Development Service (RDS) and elements of the Countryside Agency (CA). The first purpose of this new body is the promotion of nature conservation and protecting biodiversity.

---

[5] This is a slightly different definition of institutions to that used by Leach et al. (1999) who refer only to the 'rules of the game' as institutions.

[6] For an overview see Adams 1997; Fairbrass and Jordan 2003; and Bulkeley and Moll 2003.

Since October 1994 The Habitats Directive and Birds Directive have been implemented in England, Wales and Scotland through the Conservation (Natural Habitat, & c.) Regulations 1994. The regulations allow for the adoption of designated areas for conservation and the development of appropriate management agreements with land owners in order to conserve, restore or protect the site. The SSSIs are created under the Wildlife and Countryside Act of 1981. In the case of the North York Moors National Park, the SSSIs are contained within the Special Area of Conservation (SAC). The SAC covers approximately 44,000 hectares of the Park. In December 2004, the European Commission adopted the UK candidate SACs and they were formally designated on 1st April 2005. As English Nature note 'designation of an SAC is unlikely to affect the existing management of SSSIs to conserve their biodiversity' (English Nature 2005).

English Nature advises on the development of management schemes for biodiversity conservation. Agri-environmental schemes have played an important part here, including for example, Environmental Stewardship schemes, the Upland Management Scheme and Countryside Stewardship Scheme. It is thus impossible to separate biodiversity conservation regulation from agricultural support. Implementation of management schemes in the North York Moors is sensitive to people's livelihoods – shepherding/grouse shooting. This awareness ensures that regulation works in practice. Nevertheless, the most recent report to the government indicates that a significant proportion of the North York Moors National Park does not meet the national – and thus EU- conservation requirements due to overgrazing and bracken burning. The National Park Authority is itself a significant institution in the management of biodiversity conservation and links the interests of central government and regional and local organisations (see Thompson 2005).

12.4 Claims on the Landscapes of Conservation

Knowledge and the cognitive models that people constructed can be seen as an endowment which underpins perceived entitlements to the natural environment. There were three significant arenas of knowledge about the Esk catchment that stakeholder participants shared during the seminars. Scientific knowledge of the catchment was considerable and included water run-off, sediment formation and its effects, water quality, acid rain, indicator species, habitat succession, catchment characteristics and the moorland and valley landscape. Social and economic knowledge included rural development, regulation, transport, institutions – both organisations, and the rules of practice – and aesthetics. There was also substantial knowledge and understanding of the interconnections between the natural scientific and the social economic within the catchment. Knowledge covered a variety of scales, from knowledge about sediment formation in individual tributaries to understandings of the interconnections between sheep stocking rates and moorland characteristics. Distinctions were sometimes made between the sort of knowledge that locals might have as opposed to incomers or tourist visitors. Discussion in the seminars regularly focussed on what people did not know, clarifying gaps in knowledge – here, in relation to pearl mussels:

> [R1] I don't know anything about DNA work and genetics and I am trying to get things done on the ground but I know it is important, it is a useful lever for us for moving things forward, it is a justification for saying why we should do something about these [pearl mussels] here, looks good on a form. [S1; 11]

Knowledge was derived from a range of sources, those sources being important for the level of trust in the individual, or the organisation he or she was a part of, or the value that was attributed to the knowledge (Niemeyer 2004). Education, sometimes supplemented by specific training was significant for many of the expert stakeholders, but experience, particularly over long periods of time, elicited particular confidence.

> [R1] I would say for A, I would say he is..…. well respected within the Esk valley. He knows a lot of landowners and he was a shepherd before so all of the background is a good thing, very much so. He is one of a dying breed within the EA, you don't get people like that any more. There's just not the time..…. You tend not to get people that have the time to wander round the catchment. [S1; 38]

Experience was not exclusively a source of knowledge for experts: key local participants had also developed experience over many years. Some drew on personal history to evidence changes in the local ecology or economy. Observation was another source of knowledge, while others drew on historical sources. Some seminar participants were personally involved in monitoring the local environment on a non-official basis, and monitoring of flora and acid rain by local groups also came up. Personal conversations – particularly with local 'champions' – were often cited as a source of knowledge. It was gratifying that seminar participants spontaneously brought up how much they had learned during the course of the seminars.

We are already shifting into the communication of knowledge. It was notable that all seminar participants had very clear ideas of who they would turn to in order to find out anything they needed to know. For many this would be a key actor in the local community, or it could be an institution. Several of our seminar participants were identified as key actors in this sense: for example local counsellors, long-standing staff at the EA, land agents and gamekeepers. The National Park Authority was seen as a particularly helpful institution as a source of appropriate contacts with respect to any sort of expert knowledge. Locals were of the view that a few phone calls would link you to a network of individuals who would inform you of the right person to provide specific knowledge.

> [C1] You do have certain people that you know the names of that you can ring...that might be married to somebody else or a lot of people are related..…Eventually you can find things out, a few phone calls or a few asking questions. [S1; 31]

Seminar participants certainly set considerable store by locally acquired knowledge, and were of the view that trustworthy information was readily accessible within the catchment. Take the Moors Centre:

[C1] Having local people working downstairs, as well, makes a difference because you can ring up and ask them about things and they are going to know about the Moors Centre and the National Parks, but also local things that are going on, just because they live here. [S1; 31]

What factors appeared to govern the sort of knowledge that people sought out? How indeed did people select the knowledge they wanted to make use of? Two overarching factors came out of the analysis: cultural values and beliefs, and the presence of power or conflict. Goodwin (1998) suggests that it is very important for community participation in conservation that attention be paid to the collective and shared experience of place, on the grounds that people obtain a sense of personal meaning and identity from their relationship both to each other and in relation to their surroundings. Such collective enthusiasm for 'a peopled landscape' (Crouch 1992, cited in Goodwin) ran through every one of the seminars. The love of the landscape of the North York Moors was seen as a cultural value which united all who lived within the catchment, whether locals or incomers, and indeed this was also a value attributed to visitors to the area.

[DF4] It is really important, but if you live here and look out and you are surrounded by it all the time, so what – it's heather moorland.
[PR4] I think local people do appreciate it, especially in the autumn or August time, you never tire of seeing the heather or watching the sun go down.
[C4] It is the thought of a lot of other people as well, outside. [S4; 43]

Kamppinen and Walls (1999) discuss the importance of understanding 'meaning systems' or cognitive models to underpin good environmental decisions. Our participants demonstrated that the moorland landscape held multiple meanings for them as individuals, but the seminar discussion was founded on a common denominator of a 'moorland-centred' cognitive model that saw moorland as valuable in itself. It was this shared imagination and cultural heritage that underpinned a willingness to discuss more controversial and contested issues around a second cognitive model that participants also held to. This was the 'human-centred' model which saw moorland in more instrumental terms, for its recreational or its spiritual and aesthetic uses (Saarimaa 1993, cited in Kamppinen and Walls).

The playing out of power relations was correspondingly a further overarching factor in selectivity in the use of knowledge. Local understandings of the variables that conserve and sustain moorland were contrasted with the (lack of) understandings of incomers, and – more drastically – with those of some categories of visitors, such as off-road vehicle users. Saarimaa's third cognitive model – the professional model – underpinned seminar discussion of how conflict and exclusion might be managed. The professional model conceptualises moorland in terms of the model received through education, with expertise and public accountability as central ingredients. Discussion around conflicts between residents and visitors was regularly constructed in terms of national taxpayers' rights to the local moorland resource, and the role that different knowledges and understandings play in resolving such conflicts.

We have already started to identify some of the dynamics of knowledge acquisition. These are important for identifying the processes of applying knowledge to practice and a significant aspect of the process of legitimation. The seminars provided some interesting insights into possible stages in the dissemination of knowledge. From an environmental practice point of view, recognition of a problem is likely to be a first step. Participants identified problems in the course of discussion, and then – in some cases – moved on to seek explanations or understandings. A specific problem may of course be put to one side. We have already mentioned that participants regularly identified gaps in understanding, and generally had clear ideas of who to approach to gain access to required sources of knowledge. Education is an obvious process for remedying lack of knowledge and understanding of a problem, its causes or its solutions.

> [R5] It is much harder looking at all the interactions of everything, most research isn't done like that....... You can pull your science bits of research which is trying to pull in people, get them on board through meetings and site visits; I always think site visits are excellent....you walk out, you show them what's what.....say this is why we need to measure silts, for example, to prove that it is coming from here or there. [S5; 30]

This ties in with who does not know and how they might be reached.

Two further processes which are of key significance to environmental practice took place within the seminars themselves. One of these is what Folke (2003) calls 'social memory'. Folke defines this as 'the arena in which captured experience with change and successful adaptations, embedded in a deeper level of values is actualised through community debate and decision-making processes into appropriate strategies for dealing with ongoing change' (ibid. p. 2032). Others have discussed this kind of process in terms of institutional learning and memory (Olsson and Folke 2001) or, slightly differently, as 'local (ecological) knowledge' (McDonald et al. 2004) or 'traditional ecological knowledge' (Berkes 2004). What the seminar discussions also brought out was the significance of 'champions' and key local communicators for this process of social learning.

The final process that we identified was the development of 'holistic knowledge', which can usefully be distinguished from other forms of social learning. Seminar participants were continually striving to link up different categories of knowledge so as to take forward a more holistic understanding of the interconnections between their beloved landscape and the physical, social and economic factors that might change or preserve it. This contextualising process was going on throughout all five of the seminars, but a specific example came up towards of the end of the first. Several participants had learned for the first time that the Esk has a population of pearl mussels, most some 80 years old and so threatened with extinction. Participants identified the importance of SSSI designation for the river as a basis for funding to improve the pearl mussel habitat (reduced siltation), and asked the experts present to supply them with the criteria for selection and designation of SSSIs.

Our working definition of entitlements for the seminars encompassed the broad category of the rights, claims and access to the rural landscape that participants believed should be available to them. These included legal claims as well as informal rights that had arisen historically, or were a part of local tradition. The seminar explored the range of goods and services that the river catchment provided, the ways in which they depended on natural resources and the extent to which they represented entitlements that were open to all, or just to a specific group. This involved looking at how entitlements are negotiated and implemented. In particular we were concerned with the ways that people supported their claims through both local and scientific knowledges.

A range of entitlements were mentioned to us: claims related to recreation, livelihoods and access to the landscape. This last group of entitlements covered a breadth of scale from concern with the conservation of particular species to the aesthetic value of the moors landscape. There was concern for the entitlements of young people to be educated and of old people to remain within the community. The nature of entitlement claims extended from private ownership to discretion and tradition.

The debates on entitlements were dominated by issues of access in various forms, and discussion revealed issues of conflict and ambivalence and the effects of macro policy decisions on local claims. For example, access to the landscape included rights for walkers, horses, bikes and off-road vehicles as well as access to the aesthetic values prized by a number of actors.

> My perspective is that the River Esk is a natural resource for people such as photographers and painters which is a tremendous resource. The problem there is the accessibility, the wonderful part of it is the Glaisdale Woods where you have the [. . .] Way through, which gives good access without doing damage to the environment around it. [D1] [S1:11]

The walkers represented in the seminar participants were generally very satisfied with fulfilment of their claims for access.

> I think the National Park has done wonders. I mean the footpaths are well indicated, the stiles are good. . . . . .personally I don't think [Open Access] is needed so much because there are so many footpaths. I don't know, my personal view is I don't think we need a lot of access. [S1] [S1:11]

Interestingly, this quote demonstrates that claims are not unlimited, and that people were seeking a 'reasonable' entitlement. This idea of reasonable entitlement was also illustrated in the case of a local landowner continuing a tradition of allowing local people to fish and walk next to the river.

The first thing to note about perceived entitlements was their dynamic nature and the ways in which social and economic change brought about changes in actual entitlement. In some cases, legal circumstances remained the same, but the economic context changed. In others, the legal framework changed, reflecting changing social values about the environment. Let us take an example to illustrate each. The institution of private property has always made poaching an illegal claim, one that nowadays is little exercised.

> got some tremendous poaching problems. Thankfully that has changed. Closed circuit television cameras in the harbour and by-laws stopping fishing on a lot of the river and our efforts, and possibly the fact that farmed salmon is one of the cheapest things in the shops now, so there isn't the urge. A lot of the poachers have had enough of it. [A1] [S1:6]

On the other hand, changing values have led to biodiversity conservation, and the freshwater pearl mussel is now a protected species. Whereas in the past poaching of these mussels was undertaken to seek pearls, the concern nowadays is to protect the social value of biodiversity – an urgent matter when the small remaining local population is some 80 years old.

> the pearl mussels, they could be extinct in 20 years. It is our responsibility to do something now, to stop that from happening, because everything should be preserved really so that future generations can experience what we have experienced. Also everything that lives has rights in a sense. You don't want to kill things off and then that's it, they are gone for ever and will never be seen again. [C1] [S1: 37]

It is important to note here the representation of claims of non-humans and those as yet unborn (O'Neill 2001; Berry 1999).

For both salmon and freshwater mussels then, the institutional context within which the species exists has changed. The changing evolution of collective values (Goodwin 1998) was also illustrated by the contrast between wild daffodils – which used to be picked for sale in the market – and sloes. Whereas the daffodils are now protected and may only be appreciated where they grow, sloes are more widespread and are still available as a common resource for those who make their own sloe gin.

In the examples so far, personal entitlements have been discussed in the context of the regulatory institutional framework. Private property is an overwhelmingly important institution. However, the rights to use of land-based natural resources may still be very complex. The moorland, for example, is privately owned, although parts of it are common land. Danby Court Leet, established during the Middle Ages, has responsibility for managing approximately 50% of the common land of the moor: commoners have grazing rights inter alia, while the owner of the soil has mineral and shooting rights. Even on the local, small scale, the ways in which historically established institutions are worked out through the modern system has differed. In Aislaby, the Court Leet has long been defunct.

> but the freeholders of Aislaby had various, what we call garths, which are small areas of land which were let out to villagers. Those in practice, all but two of them, are not on the common land. These garths are separate from the common land. The freeholders of Aislaby have all died out. The parish council has run these garths for the lasts 12 years and we are about to stake a claim with the Land Registry for them. Things have obviously gone a different way in Aislaby. [Pt2] [S2:8]

This overview of entitlements from the seminars shows how broad a range of entitlements were recognised by participants. It also demonstrates the understanding and measured acceptance that participants had for the claims of others. If 'legitimate' means that a claim is accepted by others even when they have no immediate benefit, the tolerant attitude of different actors indicates a general legitimation of the environmental protection regulations currently in place.

Given the background of claims and the understandings that people have of the environment, we can now turn to look at the experience that people have of regulation, and in particular those policies affecting conservation. By regulation we understand formal legislation that affects behaviour in relation to the natural environment.

Regulation makes up the formal part of the institutional context. We started the final seminar by asking participants about which regulations they knew of in the River Esk catchment and how significant they are to their daily lives.[7]

The North York Moors National Park Authority and its role as a regulatory agency was – not unexpectedly – a continuing theme running through the discussion. Apart from environmental directives and their implementation, there was considerable discussion of planning regulations.[8] Specific regulatory issues arose concerning agri-environmental schemes, biodiversity conservation measures including SACs and SSSIs, access to the countryside, hunting with dogs, and the Water Resources Act.

People had both positive and negative perceptions of the regulatory framework affecting their lives within the National Park.

> the discussion of regulation seems to be focussing on regulation as a negative force – something that constrains people and no doubt it does very often constrain people, but presumably there are just as many instances where it is there actually to protect people, and certainly people living in local communities like this. There are a lot of regulations which protect their livelihoods, their environments, their way of life, and yet we don't tend to think of those. We tend to think of the ones that stop us doing things. [L4] [S4: 15]

So, on the one hand it was recognised that regulation protected the environment and associated livelihoods; on the other, that implementation in practice imposed costs.

Time and again, discussion turned to the fact that a choice had been made to preserve heather moorland as a landscape, rather than earlier juniper forest or even iron age farming. One participant was keen on the reintroduction of more wooded areas, and this lead to the following exchange:

> [C4] …that's just a whim wanting trees on the moorside.
> [P4] Don't forget that it's the regulation that governs your moorland.
> [C4] Mine?
> [P4] Your husband is part of the moorland economy. It is very strictly regulated what landowners do on there.
> [C4] Oh yes, I mean you're wanting to change it, and a lot of the tourists that come here have come to see the moors that are covered in heather like it is.
> [P4] They have been persuaded that it is beautiful. [S4; 31–32]

Such disagreement was however contained.

Running through all four seminars was agreement on the beauty of the large expanse of heather moorland. All participants and stakeholders valued this landscape very highly for themselves and for visitors – and the National Park status that protected it. Nevertheless, people mentioned the costs that this status could impose on them as local residents, from the constraints imposed by planning requirements when they want to upgrade their homes to the introduction of

---

[7] Those who attended the seminar included both representatives of agencies responsible for implementing a range of environmental regulations as well as people immediately affected.

[8] These are not investigated here as they are not immediately relevant to this chapter. It should be noted though that there is greater public participation in planning than in the conservation activities of the Park.

new styles of architecture. There was a widespread understanding that National Park designation maintains a landscape that was the product of late 18th century society.

> the whole of the moorland area is completely man-made [D4]. . . . . .regulation and funding could be accused of.. . . .stifling evolution of landscapes and rural economies and to a certain extent restricting diversity. [L4] [S4;30]

There was widespread agreement that National Park status permitted a national say in the local environment. The immediate response to our question about who should control the Park was:

> If it wasn't a National Park, it should be local people, I would suggest. But as it is a National Park, then the nation should have a voice. [P4] [S4; 39]

It was further recognised that the nation contributed to the costs of maintaining the Park through taxation, and that the National Park benefited society as a whole, through its responsibilities for conservation and public enjoyment.

Seminar participants were clear that the Park has a value beyond the national as the largest continuous area of heather moorland in Western Europe and as such is an important breeding habitat for rare birds.

> not only is it SSSI but it is also a European designated special protection area and a candidate for special conservation, which is a European regulation so it is higher up than the SSSI or anything else that we as the UK could then say. Really that designation means that it is protected for the birds, for specific species, golden plover and merlin to name two. [R4] [S4; 32]

The complexities of the interactions between scales of regulation are strikingly put by another participant:

> One moment it is this European site of special protection area, at another it is a farmer's place where he has to live and work and at the same time it is a place where someone from Leeds comes for their recreation. It is drawing all those things together and making them work. . .[D4] [S4; 42]

There are then, three groups – international, national and local – with different value perspectives and which may have different priorities with respect to regulation. Implementation may thus cause difficulties.

> We have got that situation on Slight's Moor, it is common land, but we own it. It is a SSSI and a SPA as well, but it has got ancient monuments on it that English Heritage are responsible for managing. So you have got two conflicting things there. The whole thing is subject to the CROW Act as well, so in theory you have a right to roam over all of it, but English Nature don't want people to roam all over it, because it is a site of scientific interest and a special protection area and everything else, and English Nature have got these ancient monuments on it as well. [PB]
>
> That applies to Danby Common as well, so all those rights are layered over the top of each other, and actually end up controlling or restricting or confining – or maybe supporting, to look at it in a more positive way, exactly what happens to the land. [L] [S2, 35]

Knowledge of regulations did not necessarily mean knowledge about their impact. For example, there was widespread concern, even amongst the experts, that the impact of the single farm payment system on the landscape could not be confidently foreseen. For those who were well informed about regulation, there were issues of the inconsistency between policies both within and between different scales and whether there was sufficient flexibility in the implementation and policing to accommodate these.

[L4] People want, most people would accept that you need a system where there is a certain framework for regulation, but at its point of application it has to be done in a way that can react to different circumstances. It has to be done in a logical way, a flexible way. That's difficult to manage isn't it?
[Dv4] That's where the enforcement policy comes in. Any enforcement policy has to sort of look at the range of issues and how the person who has broken the regulations has viewed it from the outset and then you look at the public interest and decide whether you are going to enforce or not.
[L4] But it is only as good as the person at the endpoint of that system in a way isn't it. It relies a lot on their experience and their relationship with the regulations that they are enforcing in the area...[S4: 15–16]

Another of the ways of overcoming these conflicts in practice is by direct payment to the landowner, but there is still some resentment that these are not decisions that local people are involved in at all.

English Nature complained to us that one of the farmers with commoners rights on the Moor was damaging the Moor near what are called the High and Low Bridestones by feeding his sheep there......It is being sorted out by this sheep/wildlife enhancement scheme by public money, by paying them to take sheep off the Moor and not feed them in the way that they are feeding them. In a way, you might say that everybody is happy. The farmer has got some money for not doing something he was doing, and English Heritage are not getting their ancient monuments damaged now, so it's taxpayers' money that is solving the problem. [PB]
......instead of being moved round the Moor, and eating mainly the heather, they [the sheep] are actually ranched and fed in big round bales......the feeding sites are green circles in the heather...... the heather won't grow, so it does have a big landscape impact. It is interesting that the ultimate control there is actually coming from completely outside the area, and the decisions and the way in which the situation is managed does not have much direct input from local people at all. [L] [S2, 36]

Although people did feel involved in local planning decisions, but there was an almost complete absence of involvement in environmental regulation. Most regarded environmental legislation as being imposed from above and outside their control.

English Nature may come and designate a particular piece of land for its wildlife value, and thereafter its use by the owners would be restricted, so that would be a restriction right. That decision about how to choose that piece of land and about how to control it subsequently is made entirely by the staff of English Nature, without any democratic process at all......English Nature as a body for example, is a quasi-autonomous government body, it is not a central government body so that control has been handed over to them......The process of restricting their [the landowner] rights on their land is not democratic. [L] [S2, 35]

As must be apparent from the quotes we have chosen, the significance of different knowledges and claims have an important role to play in participation and people's attitudes towards environmental conservation. We have covered a lot of ground, demonstrating the complexity of a multilayered situation. In order to make this clearer, we now use a case study to draw out the ways in which the process of legitimisation plays itself out in one particular situation.

12.5 Living in the Landscape: A Case Study

The aim of this case study is to look at the specific environmental objectives for the heather moorland of the North York Moors. The objective of sustainable development is illustrated by participants' discussion of cattle grids as a specific mechanism, albeit just one of many. We propose to take this mechanism to show what we can learn about the process and content of an interdisciplinary public ecology: how do local, national and international regulations to protect the environment work or not work together?

Sheep grazing is a key to the maintenance of heather moorland, but is threatened by the long term decline of time and labour for shepherding, and by flocks being taken out during the Foot and Mouth epidemic that began in 2001. If key flocks on the moors are removed then nothing will stop other sheep roaming in the spaces they have left behind thus making care of the flock much more difficult. Cattle grids have now been installed on a number of the roads that lead out of villages located in the valleys as the road goes up onto the open moorland. These grids limit sheep roaming and flocks intermingling, thus facilitating the management of sheep over wide areas. Fears were expressed that large scale ranching might still take the place of traditional sheep management despite cattle grids, perhaps as an unintended consequence of the single farm payment.

There were a wide range of environmental claims driving actors and stakeholders. Maintenance of heather moorland was seen as essential to the enjoyment of the 'natural' environment as well as for biodiversity conservation. There is a legal requirement to meet the national and European Union conservation objectives. The maintenance of livelihoods based on sheep grazing was another important claim, but the support of community was also mentioned. Other claims were more closely associated with the democratic process: that there should be transparency in decision making, and that locals should 'have a say'.

The institutional context for the construction of cattle grids was equally complex. The specific organisations named during the seminars had very different perspectives on using cattle grids as an environmental mechanism. They included the National Park Authority, the Court Leet, F&RCS (Farm and Rural Community Scheme), English Nature, the Highways Authority, Yorkshire Water and the parish councils. The rules of practice of how things get done with respect to installing cattle grids certainly depended up different knowleges on the part of actors and were more or less transparent to them. It was (agriculturally regulated) markets that provided the context for livelihoods. However, the formally based livelihoods of sheep farmers worked alongside informal ways of working together. This was either in terms of sheep farmers helping each other with critical shepherding tasks, or it was in-kind assistance from landowners who needed sheep grazing to manage and maintain the moors for shooting. Common land rights on the moors involved a complex patchwork of interests, and private residents could be affected by the noise of cattle grids. It is therefore not surprising that the funding and location of cattle grids involved complex negotiations between planning and regulatory institutions. Time was an important institution both in terms of the costs of time to sheep farmers and the time scale in terms of the FMD outbreak.

It is possible to see a series of nested scales operating with respect to cattle grids as a mechanism for maintaining heather moorland. These scales are social, spatial and ecological. The people initially and obviously affected are individual farmers, but grids work at the interface between the moor and the village. It is also problematic for shepherding that non-hefted sheep roam very widely over the moorland. Helpfully, the maintenance work of sheep farming and shooting operate at the same scale. Tourists and biodiversity conservation organisations have a desire to maintain open moorland which extends to extra-local scales. Community tastes operate at a local scale which is more informed by livelihood constraints. Indeed there is widespread concern for locally based livelihoods (including sheep farming and shooting) in the face of the extra-catchment scale of most livelihood opportunities.

Establishing the communication and limiting the conflict required to agree the location of cattle grids was rendered difficult and complex due to the need to bring together a large number of stake holders from different scales (individual sheep farmers and landowners, commoners and villager residents). For cattle grids to work in assisting sheep farmers, there was still a need for them to work together, with concern that key individuals might withdraw from keeping sheep or co-operating. Numerous stakeholders participated in the decision making around the proposals for cattle grids. While some seminar participants were adamant that information about the cattle grids proposal was widely available form Parish Councils and elsewhere, others felt that 'one morning the cattle grids started to appear everywhere'. This did however, stimulate one person to say that she thought it important that individuals seek information about changes in the local environment; one couldn't just expect it to be delivered. The breadth and depth of local knowledge about the relationship between the social and the physical (for example, sheep grazing and the maintenance of moorland) was very striking, particularly given that no farmers participated in the seminars.

Despite some reservations, the consultation process was seen to have worked well, involving a number of organisations and some residents. It was nevertheless clear that even the relatively simple action of installing cattle grids involved a lengthy process of negotiation between many organisations. It was a specific event (the outbreak of FMD) that precipitated a speeding up of the process, facilitating action. A further aspect of successful implementation is that things tend to happen when there is a 'champion' to push them forward. It is clear using just this single case study that even actions that appear to be well defined or contained and on a relatively small scale, may prove enormously complex in terms of achieving implementation and appearing as legitimate actions within the community.

12.6 Conclusions and Implications for Policy Making

> people have multiple overlapping identities that they are able to mobilise and reference groups to which they refer for legitimation and support. Citizenship is thus redefined, in effect, as engagement through emergent social solidarities, which ... are likely to be "increasingly messy and unstable" (Rayner 2003; 165).

What can we learn about the legitimation of EU policies for nature conservation from this empirical micro-level study? Our approach here has been to look at the claims that people make on the goods and services provided by the environment in the context of the formal regulations affecting nature conservation. Our assumption has been that knowledge, both formal and informal, plays a significant role in formulating those claims. As Rayner describes above, and the evidence from our empirical study shows, the situation of the individual actor is complex and dynamic.

An element of this complexity is that people do not live with a single policy, but work with multiple policies that affect the rural environment and which are implemented in a number of different ways. People are therefore living with the net outcome of the cluster of policies that affect their environment. People both make compromises with each other and through the implementation of regulations. Our discussions showed that people took responsibility for the fact that those around them needed to earn a livelihood, and that local residents, incomers and tourists had the right to enjoy the moorland landscape. What is more, people gave weight to non-human interests and the interests of citizens of the future. None of this produces a single straightforward solution to how the environment should be managed. It is a continually negotiated and messy practice.

Environmental policy affecting the moor was not negotiated. There was no process of deliberative democracy with respect to conservation in place in the moorland as a whole, only implementation through (non-compulsory) participation by individual farmers as private owners. Institutions representing environmental and economic interests on the moorland have contributed to national level development of regulations, but for the most part, individuals feel themselves at effect of decisions. People saw EU policy making as very distant. Similarly, although the British Government Select Committee system invites comments from interested parties on the development of environmental legislation, individuals felt isolated from such national level processes. However, the larger institutions, such as the Park authorities, English Nature and the Country Landowners Association did provide evidence to central government.

It is only at the micro level of specific mechanisms that participation works, and not always then. The example of the cattle grids demonstrates a participatory action to mediate the effects of agricultural and environmental policies. The grids and their placement on the moorland roads has had a successful outcome. However, this is a very long way from involvement of local people in policy formulation. There remains a danger that initiatives that stem from the grass roots will not fit with policy formed higher up either at the national or international scale, and may just encourage short-termism or waste of resources if not sustained in the longer term. This criticism has been levelled at participatory development approaches in low income countries, but needs further investigation in the context of sustainable rural development in the 'north' (see for example, Korf and Oughton 2006).

The British government requires that the management of conservation in the UK be based on 'good science'. This approach precludes the kind of embedded scientific knowledge that became evident from our empirical work. And yet it was this embedded scientific knowledge that people drew on to mediate their own claims on the environment in the light of the regulations that were in place. In other words, it is reference to a broader knowledge base that people use to legitimate action in the natural environment. Thus the public ecology approach, accepting a broader and less rigid definition of the science of the environment, provides a depth and qualitative understanding that is highly significant. On a final note, the participants in this brief study welcomed enthusiastically the series of meetings as an opportunity to increase their knowledge of the Esk valley and the moors environment. We felt this was a clear demonstration of public ecology in action.

References

Adams, W.M. (1997). 'Rationalization and conservation: ecology and the management of nature in the United Kingdom.' *Transactions of the Institute of British Geographers* 22: 277–291.

Adger, N., K. Brown, J. Fairbrass, A. Jordan, J. Paavola, S. Rosendo and G. Seyfang (2003). 'Governance for sustainability: towards a 'thick' analysis of environmental decision making.' *Environment and Planning* A (35): 1095–1110.

Berkes, F. (2004). 'Rethinking community based conservation.' *Conservation Biology* 18(3): 621–630.

Berry, T. (1999). *The Great Work: Our Way into the Future*. New York: Bell Tower.

Bulkeley, H., and A.P.J. Moll (2003). 'Participation and environmental governance: consensus, ambivalence and debate.' *Environmental Values* 12: 143–153.

Crouch, D. (1992). 'Popular culture and what we make of the rural, a case study of village allotments.' *Journal of Rural Studies* 8(3): 236–255.

Fairbrass, J, and A. Jordan (2003). 'EU environmental governance: uncomplicated and predictable policy making?' *CSERGE Working Paper*, EDM 02–03.

Folbre, N. (1994). *Who Cares for the Kids?* London: Routledge.

Folke, C. (2003). 'Freshwater for resilience: a shift in thinking.' *Philosophical Transactions of the Royal Society London* B 358: 2027–2036.

Goodwin, P. (1998). '"Hired hands" or "local voice: understandings and experience of local participation in conservation.' *Transactions of the Institute of British Geographers* NS 23: 481–499.

Kamppinen, M., and M. Walls (1999). 'Integrating biodiversity into decision making.' *Biodiversity and Conservation* 8: 7–16.

Korf, B., and E. Oughton (2006). 'Rethinking the European Countryside: can we learn from the South?' *Journal of Rural Studies* 22(3): 278–289.

Leach, M., R. Mearns and I. Scoones (1999). 'Environmental entitlements: dynamics and institutions in community-based natural resource management.' *World Development* 27(2): 225–247.

McDonald, A., S. Lane, N.E. Haycock and E.A. Chalk (2004). 'Rivers of dreams: on the gulf between theoretical and practical aspects of an upland river restoration.' *Transactions of the Institute of British Geographers* NS 29: 257–281.

Niemeyer, S. (2004). 'Deliberation in the wilderness: displacing symbolic politics.' *Environmental Politics* 13(2): 347–372.

North, D. (1990). *Institutions, Institutional Change and Economic Performance*. Cambridge: CUP.

NYMNPA, North York Moors National Park Authority (1998) North York Moors National Park–Management Plan, Helmsley.

Olsson P., and C. Folke (2001). 'Local ecological knowledge and institutional dynamics for ecosystem management: A study of Lake Racken watershed, Sweden.' *Ecosystems* 4: 85–104.

Olsson, P., C. Folke and F. Berkes (2004). 'Adaptive co-management for building resilience in socio-ecological systems.' *Environmental Management* 34(1): 75–90.

O'Neill, J. (2001). 'Representing people, representing nature, representing the world.' *Environment and Planning c: Government and Policy* 19: 483–500.

Oughton, E. and J. Birch (2004). Chapter 6.4 'Institutional Dynamics' In S. Bell *Integrated Management of European Wetlands*. Final Report, July 2004. EU Contract number EVK2-CT-2000-00081.

Rayner, S. (2003). 'Democracy in the age of assessment: reflections on the roles of expertise and democracy in public-sector decision making.' *Science and Public Policy* 30(3): 163–170.

Robertson, D.P., and R.B. Hull (2003). 'Public ecology: an environmental science and policy for global society.' *Environmental Science and Policy* 6: 339–410.

Saarimaa, R. (1993). Forester's Conceptions of the forest. MA-thesis, Department of Cultural Studies, University of Turku [in Finnish].

Sen, A. (1999). *Development as Freedom*. New York: Knopf.

Thompson, N. (2005). 'Inter-institutional relations in the governance of England's national parks: A governmentality perspective.' *Journal of Rural Studies* 21: 323–334.

Chapter 13
Towards Governance
and Procedural Legitimacy?

Marielle van der Zouwen

13.1 Introduction

Oughton and Wheelock's chapter 'Conservation in context – A view from below' provides us with an interesting insight into the perceptions of people involved in conservation practices. Why is this interesting? First of all, the authors give us an insight into the practice of nature conservation policy and especially into the diversity of perceptions of the people involved. Secondly, the authors state that the institutional context affects the actions of actors in conservation policies. For the UK with its long tradition of nature conservation activities, this is a relevant starting point. Many contemporary policy practices bear traces of initiatives from the late 19th and early 20th centuries.

The empirical richness of Oughton and Wheelock's study really highlights the trends that are at the core of this volume: the shift from government to multi-level governance, on the one hand, and the growing importance of procedural legitimacy, on the other. I am convinced that we need an in-depth study that unravels the emergence and development of such trends to understand what happens in policy practice. Based on Oughton and Wheelock's study, I would like to address the questions of to what extent and how the trends mentioned above manifest themselves in the North York Moors case. To answer these questions, we first need a more explicit description of what these two trends encompass. I will do so by formulating expectations for each trend. Consequently, I will address to what extent these expectations are met in the North York Moors case.

13.2 Expectations

Regarding a shift to multi-level governance, I formulated expectations for four dimensions: discourses, actors and coalitions, power and resources, and rules of

Marielle van der Zouwen
Forest and Nature Conservation Policy Group, Department of Environmental Sciences,
Wageningen University, Netherlands
E-mail: marielle.vanderzouwen@wur.nl

J. Keulartz and G. Leistra (eds.), *Legitimacy in European Nature Conservation Policy:*
Case Studies in Multilevel Governance, 177–181. © Springer 2008

Table 13.1 Expectations on multi-level governance for four dimensions

Discourses	Emergence of multiple challenging discourses from various levels
Actors and coalitions	Emergence of coalitions of a variety of actors from different levels
Power and resources	Emergence of interdependencies among actors, no real dominant actor, facilitating role of governmental actors, mobilisation of sub-national and supra-national resources
Rules of the game	Emergence of goal seeking processes, trust, ad-hoc coalitions and decision making, increasing importance of sub-national and supranational rules

Table 13.2 Expectations on legitimacy issues

Legitimacy sources	Decreasing use of substantive legitimacy sources – e.g. scientific expertise, status – less important
Input legitimacy	Increasing inclusion of different views in agenda setting and decision-making
Output legitimacy	Increasing recognition of different views in policy products
Throughput legitimacy	Increasing importance of procedural rules

the game. These dimensions are derived from the policy arrangement approach (Van Tatenhove et al. 2000; Arts and Leroy 2006). This approach focuses on the temporary stabilisation of the organisation and substance of a policy domain in which temporary stabilisation is conceptualised as a 'policy arrangement'. The content concerns *discourses* on relevant issues advocated by actors involved. The organisation is operationalised in terms of the dimensions: *actors and coalitions*, *power and resources*, and *rules of the game*. It is not my intention to reconstruct policy arrangements in this contribution. Rather, I use the four dimensions to formulate more explicit expectations for the trend of multi-level governance. Using these four dimensions enables a nuanced answer to the central question of this response to Oughton and Wheelock (for a more detailed explanation and application, see Van der Zouwen 2006). This results in the following four expectations (Table 13.1):

In addition to the four multi-level governance expectations, I have also formulated four expectations on legitimacy. In so doing, I followed Engelen et al. (this volume, pp. 3–21), who, mainly drawing on Scharpf (1999; 2004), state that procedural legitimacy has gained more prominence, whereas the legitimating power of substantive sources has decreased. In their introduction Engelen et al. distinguish between input, output and throughput legitimacy. This results in four expectations on legitimacy issues (Table 13.2):

13.3 Towards Multi-level Governance?

As far as *discourses* are concerned, the North York Moors case shows the presence of a broad range of different views on nature and land uses by various people involved in conservation processes. Authors that have studied nature policy in the UK have demonstrated the presence of a so-called 'great divide' in this policy field: historically, there is a strongly institutionalised separation between actors who focus on the conservation of the aesthetic appearance of the landscape and its public enjoyment, on the one hand, and actors who focus on the conservation of species and habitats, on the other (e.g. MacEwen and MacEwen, 1982; Winter 1996). Up till the present day, this divide has been represented in the dominance of specific organisations, each of which has its own view on nature and accompanying rules, such as the National Park Authority and English Nature. Although Oughton and Wheelock stress the importance of the institutional context for contemporary nature policy, their study does not show whether the great divide is relevant for North York Moors' conservation policy or which discourse is dominant As a study on dynamics in nature policy in another English national park has shown that the divide still heavily impacts on current policy practices (Yorkshire Dales; Van der Zouwen 2006), it could be expected that this might also apply to the North York Moors case. Furthermore, it is also unclear what sort of role is played by discourses from the international level such as 'sustainable development' and 'agricultural modernisation' stemming from the Common Agricultural Policy.

As far as the *actors and coalitions* are concerned, the North York Moors case shows that conservation is a topic in which both governmental and non-governmental actors from different levels are engaged. As the case does not address the involvement of EU organisations such as the European Commission or the European Court of Justice, it seems that all the actors involved are of domestic origin. The multi-level involvement of actors is an internal UK affair. It is hard to say whether this also goes for the formation of coalitions. The study did not address which coalitions between which actors emerged in the area.

Regarding *power and resources*, Oughton and Wheelock state that 'private property is an overwhelmingly important institution.' As in other national parks in England, a large part of the park area is in private hands, and consequently, it would be interesting to see to what extent the structural power of private landowners, in particular farmers, plays a role in conservation policies. The study shows that people do perceive the existence of this type of power. However, it is unclear how this impacts the positions of those involved in conservation policies. In this respect, we have to keep in mind that private landowners rarely participated in the themed meetings that were organised by the authors. Consequently, the study does not reflect the landowners' perceptions. Next to private property, the strong position of governmental actors such as the National Park Authority and English Nature is also characteristic for English national parks. This is a stable factor in national park conservation policies partly due to the strong institutionalisation of the great divide. It would be worthwhile investigating whether the pivotal position of these governmental bodies is also present in the North York Moors. In particular, the position of the Court Leet as an example of a traditional actor would be an interesting subject for further study. And what about the mobilisation of EU resources and interactions between the regional and the EU level? To what extent is the mobilisation of EU funding enabling or constraining new coalitions in the national park? And, as some multi-level governance experts claim (see, for instance, Marks et al. 1996), do we see regional actors (the National Park Authority) interacting directly with the EU level (European Commission) and thus bypassing the national level?

As regards *rules*, Oughton and Wheelock demonstrate that many regulations from different levels are perceived as playing a role in the daily practices of the park's conservation policies. At the same time, some questions on governance rules remain unanswered, such as: is the process of an ad-hoc nature and is it focused around a specific project; to what extent is there room for local traditions and for EU and global rules on nature conservation; and what is the role of trust among actors?

At first sight, the authors' detailed study indicates that various levels are involved in each of the four dimensions. At the same time, however, we only witness a partial shift from government to multi-level governance. It seems that many traditional characteristics still impact today's conservation policies in the area. Consequently, present day conservation practices show complex mixtures of 'old' and 'new' conservation policies.

13.4 Towards Procedural Legitimacy Sources?

Concerning legitimacy issues, we can also witness these complex mixtures of traditional and new characteristics. Indeed, Oughton and Wheelock show that not just *substantive legitimacy sources* are considered important. Nevertheless, given the perceived importance of actors, such as English Nature, which largely underpin their work with natural scientific research, one might expect that scientific knowledge is still a legitimate and important source for nature policy in the area and will remain as such in the (near) future. Furthermore, many nature protection designations in the North York Moors are based on ecologically defined standards (Sites of Special Scientific Interest, a Special Protection Area under the Birds Directive and a Special Area of Conservation under the Habitats Directive).

With regard to *input legitimacy*, the authors demonstrate the presence of different claims and 'entitlements'. We do not know, however, if and how different actors are actually present in agenda setting and policy-formulation processes in the North York Moors case. This is especially relevant since the role of governmental actors in the National Park Authority is traditionally strong. Thus, the inclusion of other actors in the decision-making process is not necessarily rooted in policy processes in English national parks. To determine whether we could really witness a shift towards input legitimacy, it would be interesting to know what the actual status of the themed meetings is and to what extent the input during these meetings has been incorporated into the park's agenda-setting and policy-formulation processes.

As for *output legitimacy*, within the scope of a specific project, we can recognise different views in the cattle grid debate. At the same time, we cannot distil from the study whether the outcomes and results of the debate meet the objectives of the project and reflect the views and wishes of the actors involved (most importantly because farmers did not take part in the seminars organised by the researchers/authors). If we assume that the outcomes have largely satisfied the actors involved, it is then interesting to note that an 'external event' (foot-and-mouth disease) finally facilitated the cattle grid process. Would a satisfying outcome have also been achieved without this trigger?

Regarding *throughput legitimacy*, the cattle grid project indicates the existence of transparent procedures – such as the availability of information on the proposals – for some actors, while others perceived the appearance of the grids as a surprise ('one morning the cattle grids started to appear everywhere'). It seems thus that some sort of a participatory process was set up, although there are also indications that this process was not easily accessible for everyone.

13.5 Conclusion

In the previous sections I tried to analyse the results of Oughton and Wheelock's study on people's claims and entitlements in terms of the emergence of a shift from government to multi-level governance and the increasing importance of non-substantive legitimacy sources. For both trends I came to the conclusion that traditional and new characteristics stand side-by-side. How these old and new features are (inter)related is a question that can only be answered by studying these trends in specific cases. If it were our aim to understand the character of nature policy processes and the impact of and relationship between trends like multi-level governance and different forms of legitimacy, investing in the 'power of example' would be a promising way forward. The study of Oughton and Wheelock serves as an interesting example of a detailed study on people's perceptions. Hopefully, more authors will follow.

References

Arts, B., and P. Leroy, eds (2006). Institutional Dynamics in Environmental Governance. Dordrecht: Springer.

MacEwen, A., and M. MacEwen (1982). *National Parks: Conservation or Cosmetics?* London: Allen and Unwin.

Marks, G., F. Nielsen, L. Ray and J. Salk (1996). 'Competencies, Cracks and Conflicts: Regional Mobilization in the European Union'. In G. Marks, P. Schmitter and Streeck, eds. *Governance in the European Union*. London: Sage, pp. 164–192.

Winter, M. (1996). *Rural Politics, Policies for Agriculture, Forestry and the Environment*. London: Routledge.

Scharpf, F. (1999). *Governing in Europe. Effective and Democratic?* Oxford: Oxford University Press.

Scharpf, F. (2004). 'Legitimationskonzepte Jenseits des Nationalstaats.' *MPIfG Working Paper*, 35. Cologne.

Tatenhove, J. van, B. Arts and P. Leroy, eds (2000). *Political Modernisation and the Environment. The Renewal of Environmental Policy Arrangements*. Dordrecht: Kluwer Academic Publishers.

Zouwen, M. van der (2006). *Nature Policy between Trends and Traditions – Dynamics in Nature Policy Arrangements in the Yorkshire Dales, Doñana and the Veluwe*. Delft: Eburon.

Part IV
Countries

Belgian Landscape (by Kris Decleer)

Chapter 14
Endangered Legitimacy

Nature Policy in Flanders: The Sloppy Management of Participation

Dirk Bogaert and Pieter Leroy

14.1 Introduction

Since the regionalisation of Belgium in the 1970s allocated almost all competencies over nature policies to Flanders, Flemish nature policy has developed rapidly. Particularly from the 1980s onwards and thus in a relatively short time span, it has turned into a well-established policy domain with its own legislation and regulation, civil service, scientific institute, policy plan and advisory bodies. However, as nature policy-makers work out their own environmental standards and conditions and their concomitant spatial claims, it increasingly collides with adjacent sectors and policy domains. In particular, the consecutive collisions with agriculture and with spatial planning have substantially slowed down the actual implementation of Flemish nature policy ambitions.

One of the most interesting and illustrative cases of such a conflict over claims and competencies is the implementation in Flanders of the nature policy concept that, from the early 1990s onwards, has become an important issue throughout Europe: the creation of ecological networks, also known as Natura 2000. The first Flemish policies under Natura 2000, the so-called Green Main Structure (GMS), failed due to a lack of societal and political support in the mid 1990s. As Europe expected Flanders to take part in Natura 2000, the aim of an ecological network remained on the Flemish agenda. Hence, the Flemish Ecological Network (FEN) that was launched after the failure of its precursor, was in fact a sort of replica of the former initiative. On the basis of the lessons learned from the GMS policy disaster, however, a uniform policy implementation and a clear communication were identified as crucial for acquiring legitimacy. One would therefore have expected new initiatives to incorporate these lessons. As the empirical part of this article will demonstrate,

D. Bogaert
University of Amsterdam, Department of Geography, Planning
and International Development Studies
E-mail: eengelen@fmg.uva.nl

J. Keulartz and G. Leistra (eds.), *Legitimacy in European Nature Conservation Policy:*
Case Studies in Multilevel Governance, 185–204. © Springer 2008

however, the opposite was the case. Due to bad communication and messy strategies on public participation, the FEN also nearly failed. It's ultimate implementation came after huge delays and only after the initial ambitions were substantially restricted.

This article first sketches the institutional development of Flemish nature policies, in order to give the wider context for the GMS failure (section 14.1). A brief intermezzo presents the developments in Flemish nature policies immediately after the GMS failure, focusing in particular on the increasing multi-actor, multi-sector and multi-level features of Flemish nature policy making and the difficulties these imply (section 14.2). We then move on to the FEN issue, enumerating the lessons drawn from GMS, listing the renewed ambitions of FEN, evoking the political and societal battles that it incited, which were reinforced by a messy strategy of public participation (section 14.3). We conclude by pointing out that the strategies for public participation in particular were a main source of obstruction of the FEN initiative, and have damaged the legitimacy of nature policy in general.

14.2 A Brief History of Flemish Nature Policies

14.2.1 Early Stages

Private nature protection has existed for quite some time in Flanders, but interference by the state is of a more recent date (Van der Windt 1995; Van Hoorick 2000). Flanders had to wait until the last decades of the 20th century before a more or less systematic government policy regarding protection of nature and landscape came into being (Bogaert 2004). As elsewhere in Northwest Europe, the issues of nature and environment received an important impulse from the European Nature Conservation Year in 1970. This initiative led to the proposed 1973 Law on nature conservation, aiming at a comprehensive conservation of both species and areas and providing the legal framework for no less than 30 implementation decrees (Van Hoorick 2000). These implementation decrees, however, did not come about, nor did the Law on nature conservation. The political obstacles proved too large and the political window of opportunity closed. It was not until 1981 that a royal decree settled the recognition and financing of nature reserve areas as well as the conservation associations managing these sites.

A coalition of private associations and a number of affiliated scientists insisted on implementation of the Law on nature conservation. It took untill 1984 and 1992 before two further regulations building upon this Law were announced: the *Bermbesluit*, the decree on the roadside flora, and the *Vegetatiebesluit*, the decree on vegetation. Although the Law slowly went into effect, people did become aware that the Law on nature conservation could do more than just delay the actual implementation of nature policy. International policy developments, new scientific insights and their implementation in Belgium and Flanders made it clear that nature policy needed a new, more adequate legislative basis. From the mid 1980s scientists as well as advocates from nature conservation organisations pleaded for a more adequate regulatory framework (Kuijken 1988; Celen et al. 1990; Verheeke 1990). This plea would eventually result in the Flemish Nature Decree of the late 1990s.

The step-by-step transformation of Belgium into a true federal state had serious repercussions for nature policy making. As of 1980, the competencies over nature policy making have nearly completely been devolved from the state to the regions. As a result, Flanders henceforth has established its own ministry of the environment (including nature), its own Administration for Spatial Planning and the Environment (AROL), later on renamed the Administration for Environment, Nature, Land and Water Management (AMINAL), its own advisory bodies (e.g. the Flemish Council for the Environment, later the Environment and Nature Council for Flanders; the Flemish High Forest Council; the Flemish High Council for Nature Conservation) and its own scientific Institute for Nature Conservation. Regionalisation garnered more governmental capacity for nature policy. Another effect of regionalisation was an important shift in the division of power over nature policies between nature conservation interests and agricultural interests. The latter remained almost completely nationally organized. Only in 2001 was that competence too devolved to the regions.

In short, from 1970 till 1989 Flemish nature policy gradually developed into a formal policy domain. It got its own legislative and policy-making infrastructure, including a scientific infrastructure, bodies for advice and consultation, as well as administrative capacity for the implementation of these policies. Moreover, it was organised as a separate policy domain, following fashionable concepts of the time that conceived nature policy as a separate sector. The idea that nature conservation and nature development would (have to) be linked to other, adjacent policy domains was hardly expressed at that time. Nature policy in Flanders thus primarily consisted of nature conservation in a strict sense. Protecting species and areas prevailed over a more comprehensive vision on the integral quality of the environment that came *en vogue* later.

14.2.2 The Green Main Structure: A Discursive Innovation to be Implemented

As of the mid 1980s an important discursive innovation took place in Flemish nature policy in response to new international concepts, such as 'ecological networks', 'nature development' and 'ecological restoration'. Three mechanisms were at the basis of this innovation, provoking, in turn, a more proactive approach to nature conservation (Bogaert 2004). First, conservationists realised that the small size and the highly fragmented character of the Flemish nature values did not offer sufficient guarantees for effective nature conservation. Therefore they envisaged the development of a more encompassing approach to conservation policy. Secondly, while developing that more proactive and comprehensive strategy, conservationists succeeded in linking their concerns to new scientific insights from the international nature conservation community. Based on the island theory, which emphasized the importance of migration possibilities for species, and on the importance of the proximity and distance of other nature areas, the international nature conservationists argued for a network of nature areas. Only such a network of areas, albeit with different protection regimes but linked by landscape elements that could act as corridors, could stop fragmentation. These visions were voiced all over Europe. The goal became the construction of so-called 'ecological networks', both at the level of each country as at a (pan-)European level (Kuijken 1996; Van der Windt 1995; De Jong 1999; Van Hoorick 2000; Bogaert 2004; Pinton et al. 2005). This vision was the third mechanism of innovation in the nature policy discourse: the development and implementation of new policy strategies at international level, which resulted in the EU Biodiversity Treaty and the Habitat Directive (both from 1992). In Flanders, these three mechanisms furthered the policy strategy known as the Green Main Structure (GMS).

Before introducing the GMS as such, we must point out a series of new initiatives in Flemish nature policy undertaken the early 1990s. In a short period of time, the Municipal Nature Development Plans (GNOPs), the Ecological Action Areas (EIG) and the Regional Landscapes (RL) were introduced as the main instruments and pillars of a new nature policy. These innovations were rapidly and successfully launched, albeit in a rather uncoordinated and haphazard way. Their implementation required the active involvement of local authorities and other agents, coming from sectors such as agriculture and recreation, which were not accustomed to nature policy involvement, in particular in Regional Landscapes. While garnering much enthusiasm and generating creative forms of collaborations, that were assumed to be co-operative undertakings of representatives from public bodies, from market parties and from civil society organisations, which were deemed to be exemplary for new modes of governance, these new policy initiatives also revealed some drawbacks:

- They did not have a (fully) formal legal foundation but displayed an abundance of informal rulings. This informality led to creativity, but it also contributed to a growing legal uncertainty for the parties involved.
- Local authorities were, somewhat to their surprise, actively involved in nature policy making. This did result in positive developments, but also raised questions and problems regarding expertise, capacity and, therefore, legitimacy.
- Other actors, such as farmers, entrepreneurs from the hospitality branches, hotels/restaurants/cafes, those engaged in the preservation of the cultural heritage and others also became actively involved in nature policy. This resulted in the emergence of new, divergent and probably conflicting ideas on nature, but it also gave rise to new forms of interaction, organisation and processing. As a policy domain traditionally dominated by governmental initiatives and bodies, established interests had a hard time accommodating these newcomers.

In short, these new initiatives rapidly generated more sympathy for nature policy making, but their unsystematic preparation and unclear status also caused irritation, frustration, protest and legitimacy problems among other, newly involved actors. It is against this background that we have to understand the conflicts with the agricultural sector and the social unrest the GMS produced. A related issue is the European debate on the future of agriculture with the GATT negotiations as a focal point. The third factor to take into account is the manure problem in Flanders. Strict European regulation on acidification and eutrophication, developed at that time, required a new policy approach to manure production at the Flemish level, since the extensive cattle breeding industry in Flanders produced huge manure volumes and pollution. In January 1991, the Flemish Parliament approved a decree to protect the environment against pollution by fertilizers, which became effective on 1st March 1991. Eco-activists and nature conservationists deemed this so-called Manure Decree and the concomitant establishment of the Manure Bank to be insufficient. The agricultural sector, on the other hand, perceived it as a direct assault on its economic future. During the 1990s, the Manure Decree and its gradual implementation in so-called Manure Action Plans (MAPs) turned into salient political controversies and focal points for the expressions of wider social discontent. During these debates and battles, agricultural organisations and environment and nature organisations were often flatly opposed to political parties (Wiering 1999).

Given this context, it is hard to underestimate the controversial character of the GMS. The GMS was mainly the construct of the private nature conservation organisation *Natuurreservaten*, the current *Natuurpunt*, and of some scientists affiliated with this organisation. The idea was adopted by the Flemish government in the first Flemish policy plans: the MINA-2000 plan (1989) and the more extensive Environmental Policy and Nature Development Plan (1990). In adopting these plans, the government, for the first time in Flemish history, formulated clear allocation objectives for nature conservation: 72,000 hectares before 1995, which would be better protected than only by their designation as a *nature area* in spatial planning. In order to realise this, the initial maps foresaw different categories of GMS-areas. Apart from miscommunication about the status of the maps on which the GMS protection areas were designated, the GMS became controversial for mainly two reasons. First, the links between the GMS and the MAP, the 'nature core areas' of GMS, most of which were also Bird- and Habitat Directive Areas, were supposed to be manure free areas, while the other GMS areas were supposed to entail increasingly firm restrictions on the spreading of manure. Second, although neither the GMS nor the maps on which the GMS protection areas were based had any formal status, the administration had already used them for assessing concrete permit applications. As a consequence, farmers protested the GMS as they contested the MAP, the former being regarded as a continuation of the latter by other means. In the perception of the farmers, nature conservation policy was regarded as anti-agricultural policy in disguise. In brief, the debate over the GMS had started and the protest against it was already largely organised before the minister could even formally start parliamentary discussion on the proposal.

While this societal mobilisation unrolled according to its own logic, it simultaneously undermined the process of consultation and advice at the regional (Flemish) and, in particular, at the local levels (De Blust et al. 1995). Unfortunately, the consultation and participation procedure embedded within the GMS paralleled the protest against the MAP, which in September 1993 reached its apogee with thousands of farmers on the streets, blocking traffic, and causing havoc. The hearings on the GMS were an easy target; both civil servants and nature conservationists who attended these had a hard time because of the farmers' violent protests, which at times even became physical. As the minister for the environment could not guarantee the safety of his personnel, he decided to terminate the consultation rounds. The increasing polarisation between farmers and nature conservationists initially resulted in the delay of the GMS. Further decision-making on its design and implementation was postponed until the Flanders Structure Plan (FSP), Flanders' regional spatial plan, would be determined. The latter indicates that policy makers expected the FSP to be able to help reconciling the spatial claims of nature policy and the agricultural sector. Later, it became clear that the GMS had silently been cancelled.

From the foregoing, it is clear that the controversy over and the eventual withdrawal of the GMS initiative was partly a matter of the coincidental crisscrossing of the policy processes of GMS and MAP. Secondly, the events illustrate that the advocates of Flemish nature policy had been insufficiently prepared; they rather naïvely assumed to possess strong political support, as, for instance, is indicated by the ill-timed publication of the maps. From a more institutional point of view, the conflict shows that the political agents within the domain of nature policy, which had developed as an autonomous policy domain, were badly prepared for a more comprehensive approach, controversies with adjacent policy fields and competing spatial claims, and for open and participatory policy-making. The agricultural sector's protest, which was largely supported by the Christian-democratic party (then CVP, now CD&V), played a major role in the premature resignation of the Flemish government.

14.3 Nature Policies in the 1990s: Towards Multi-actor and Multi-level Governance?

The initiative of the GMS failed not only because it suffered from faulty preparation but also because it faced firm societal and political opposition, mainly from the agricultural sector. The realisation of a Flemish ecological network, however, was to remain on the Flemish agenda, if only because Europe expected Flanders to take part in Natura 2000. Hence, the launching of the Flemish Ecological Network (FEN) in the late 1990s as a sort of replica of the GMS initiative. Below, we describe the processes of communication and implementation, as well as the participation strategy of the FEN, as they were envisaged and carried out in practice.

However, before doing that, we would like to describe briefly the changes in nature conservation policies in Flanders in the 1990s as a sort background. In particular, we want to address the growing importance of Europe, on the one hand, and the increasing involvement of a series of stakeholders and adjacent policy domains, particularly from spatial planning, on the other. While nature policy has originally been institutionalised as a predominantly state and NGO dominated domain, it increasingly gained features of what has been described as multi-actor, multi-sector and multi-level governance (Marks et al. 1996; Van Kersbergen and Van Waarden 2001; Bogaert 2004). The 'classical', bilateral management of nature policy, largely monopolised by the elites of nature conservation organisations and governmental bodies, has gradually been replaced by other forms of management (compare Pierre 2000).

The growing importance of Europe originated from the publication of directives, e.g. the Birds and Habitat Directives, in the early nineties. From that moment onward, the European Court of Justice played a crucial role in the process of further Europeanisation of nature policy making, by ruling on appeals over (the lack of) compliance of member states (Cliquet et al. 2005). As has been known from other states, Belgian or Flemish nature conservation organisations too took legal action against Flanders as it did not comply, complied too late or insufficiently with its EU obligations. In the Netherlands, Dutch nature organisations followed a similar strategy (Van der Zouwen and Van Tatenhove 2002; Bogaert and Gersie 2006), forging thus a de facto coalition of national or even local nature conservation groups with European bodies. As a consequence, hitherto relatively autonomous nation-states or, in the Flemish case, regions, unhappily found themselves stuck between top–down and bottom–up initiatives. The process of Europeanisation was also driven by another mechanism. In some cases, for instance with the LIFE funds, Europe directly supported nature conservation projects in the region, surpassing the involvement of the member state. These examples show how nature policy is increasingly developing into a multi-level, a multi-sector and a multi-actor policy domain.

All this did not imply that the role of the Flemish government was over. As we mentioned above, the lack of a formal legal foundation for the renewed nature policy led to growing management, implementation and legitimacy problems. From 1995 onwards, there have been at least three important regional initiatives to solve these shortcomings. First, there was the Decree on the General Provisions on Environmental Policy (DABM) from 1995. This decree prescribed a cycle of reporting on the state of the environment and nature, in relation to documents and procedures for policy planning. The Environmental Policy Plan, a 'white paper', would also include so-called 'binding stipulations', which would, in turn, provide the legal grounds for encompassing policy strategies such as the GMS (and later the FEN). Second, the Flemish government published the Decree on Spatial Planning (1996), which delivered the legal ground for Flanders' so-called Structure Plan (FSP), published in 1997. Thirdly, the government decreed a 'Law regarding nature conservation and the natural environment' (1997). This latter piece of legislation, called 'The Nature Decree', furthered and strengthened the ongoing institutionalisation of Flemish nature policy, including its financial, personnel and expertise resources, on one hand, and provided the legal foundations for nature development as a policy strategy on the other.

In a few years time, these three initiatives have provided Flemish policy making with a new legal framework on environmental quality, spatial planning and nature policies. Unfortunately, there were strong legal incompatibilities between the three initiatives, despite their shared temporal origin, causing legal insecurities, economic uncertainties and political unrest among several (new) stakeholders. We give two brief examples while pointing at some causes and some effects of this lack of co-ordination.

The Nature Decree is considered a compromise between (the interests of) nature conservation and agriculture. Due to the conflicts over the GMS, the responsible minister spent a large amount of time on getting input from the agricultural sector in all stages of the new decree's design. This onesided co-design, however, went at the expense of other actors that were partially new to nature policy, e.g. land owners, inhabitants of weekend houses, hunters, fishermen, the tourist sector, etc. The discussions with the agricultural sector on the Nature Decree and, inevitably, on the principles and spatial repercussions of the FEN-strategy mainly focused on the decree's allocation objectives (hectares for nature), its criteria for delimitation (location targets), its restrictions of the use of protection regimes (nature only versus joint use) and its compensating measures. By that time, in the late 1990s, the GMS concept, being 'politically burnt', had already been put aside and replaced by the 'Flemish Ecological Network' (FEN). Compared to its predecessor there are substantial differences. First, the Nature Decree introduced some new concepts, for instance, Large Units of Nature (GEN) and Large Units of Nature in Development (GENO), which accentuated the developmental character of the strategy. Secondly, the Nature Decree proposed less ambitious allocation objectives; while the indicative maps of the early 1990s designated about 377,000 ha. as GMS the new decree designated only 125,000 ha. as FEN and another 150,000 ha. as 'nature acquisition area' (IVON). These 275,000 ha. had to be delimited within five years after the Nature Decree came into force, hence in 2002. This decrease in the envisaged and claimed surface for nature conservation clearly results from the conflicts and the subsequent negotiations with the agricultural sector. That bilateral agreement, however, failed to consider the claims and interests of other stakeholders.

The second example is derived from the incompatibilities between the Decree on Spatial Planning and the Nature Decree. As stated, the former Decree rapidly followed the latter, and one would hence have expected the two decrees to build on one another. Indeed, the Flemish government and Flemish parliament committed themselves in the Spatial Planning Decree to designate 125,000 ha. Large Units of Nature (GEN) and Large Units of Nature in Development (GENO) and 150,000 ha. nature acquisition areas (IVON) in 2007 (sic!), which was clearly after the deadline laid down in the Nature Decree (2002). Furthermore, the Flanders Structure Plan provided 38,000 ha. additional nature areas and 10,000 ha. areas for 'ecologically friendly forest extension', which were to be determined in the regional spatial plans. Apart from the different deadlines, spatial planning policies also introduced a slightly different terminology and different criteria of designation than did nature conservation policy documents.

From the very start, this lack of compatibility between both policy domains created confusion, legal insecurity and legitimacy problems. At the time the Flanders Structure Plan was to be approved, social unrest about its consequences had already developed because the plan's publication was matched by a more strict enforcement policy against violations of spatial regulations in general, including some spectacularly filmed demolitions of illegally built villas and weekend houses. In the ensuing public debate, legal uncertainty about spatial policies, building permits and violations developed into a sore spot. Public debate associated 'rules' and 'restrictions' with 'nature', and some interest groups and right wing political parties were eager to abuse this association, linking nature to etatism, state control etc., which was said to be incompatible with 'libertine Flemish mores' concerning space and landscape.

The dispute mobilised a new group of opponents against nature policies: the inhabitants of weekend houses, the owners or inhabitants of 'homes located in areas not designated as residential', landowners and others. As was the case with the linking of MAP and GMS, nature policies again suffered from controversial measures in an adjacent policy domain, which was detrimental to its legitimacy. Whether this link is 'justified' by 'facts' or not, is far less important for the analysis than the observation that the link was made and that it negatively influenced nature policy. Once again, the agents involved in nature policy making appeared to be unprepared to deal with 'borderline conflicts' with adjacent policy domains or, more generally, to acknowledge the multi-actor and multi-level context in which they were situated.

14.4 The Flemish Ecological Network (FEN): A Case of Endangered Legitimacy

In this section we describe in some detail the process of communication and implementation, and the participation strategy for the FEN as they were envisaged and carried out in practice in 2002–2003 (for a more detailed description and analysis, see Leroy and Bogaert 2004). In particular, we emphasize the extent to which the handling and processing of the FEN have contributed to its acceptance or rejection, and to the legitimacy of nature policy or its decrease in general. Apart from our general research on the Flemish case, three particular empirical investigations have been used in the following analysis. Two of the investigations have analysed the implementation of different nature policy projects in Flanders from a participation and legitimacy perspective (Bogaert 2004; Leroy et al. 2004). The third research project concerns a content analysis of Flemish newspapers over the period 2001–2003, which displays the main patterns in the social and political reactions to the FEN, in relation to the overall policy development (De Zitter et al. 2003).

14.4.1 The Policy Process Concerning the FEN: Lessons Learned from the GMS?

Given the previous experiences with the GMS, one would have expected a deliberate process of policy implementation and a well thought-out communication strategy to back up the FEN initiative. Indeed, the government and in particular the minister responsible for the environment took the failure of the GMS to heart and used it as an impulse to create an all encompassing plan for communication about and participation in the FEN. This plan anticipated different ways of deliberation with different parties about the new nature policy during its consecutive stages.

As a first step, an intersectoral working group was established to steer the process of designating and acquiring FEN areas. This so-called VERAF group (*'VERwerving en AFbakening'*: acquisition and delimitation) included representatives from the Nature Department (*Afdeling Natuur*), the Forest & Green Department (*Afdeling*

Bos & Groen), the Administration of Spatial Planning (*Administratie Ruimtelijke Planning*) and the Flemish Land Company (*Vlaamse Landmaatschappij*). In practice, the deliberations within this group went far beyond acquisition and delimitation and boiled down to strategic discussions on the goals and management of Flemish rural areas as a whole. Secondly, and related with these intergovernmental deliberations, informal deliberations were held between key societal actors. These actors represented interests from nature (the NGO *Natuurpunt*), forest (the large forest owners) and agriculture (*Boerenbond* and *Algemeen Boerensyndicaat*) in the so-called Klankbordgroep ('feedback group'). The latter initiative was, in fact, a continuation of the practices of deliberations on nature policy between government and a (limited) number of actors discussed earlier. Other actors, such as small landowners, hunters and fishermen, the tourist sector etc., remained excluded. These latter actors, not surprisingly, experienced the policy process regarding the FEN as, once again, a closed and intransparent process. To conclude, the selection of who was to participate was neither clearly motivated nor acceptable to those excluded, many of whom would clearly be affected by the FEN. Their exclusion at this stage of the FEN process would, as we will see, cause huge problems later on.

Over time it became clear that the Flemish government intended the designation of FEN areas to take place in two stages. The first stage would only apply to so-called 'green areas', meaning areas already 'greened' according to spatial plans, and to so-called 'consensus areas', which would not result in disagreements. The latter were selected by comparing the map of spatial claims of the so-called desired agrarian structure (*'GAS'*) with those from the Nature and Forest Priority Map (*'PNB'*). By limiting the first stage to these *'green areas'* and *'consensus areas'*, the initiators hoped to be able to report rapid successes and avoid highly politicized location discussions. This proved to be a miscalculation. Simultaneously, the maps and measures provided for the respective areas would be discussed with the representatives of nature conservation, agriculture and forestry sectors. Based upon these discussions and negotiations, the Flemish government would determine the first selection of FEN areas by 19 July 2002. However, this was to be but a preliminary selection. After a series of modifications to the selection, the Flemish government finally approved the delimitation of the first stage of FEN, including accompanying decisions on restrictive measures and compensations for them, on 19 September 2003, more than a year after the initial selection. At this stage the areas consisted of 86,700 ha. This meant that the 'second stage' – on which decision process are still due (spring 2007) – should comprise another 38,000 ha. 'new' green areas in order to reach the target of 125,000 ha.

Apart from the organisation of negotiations within the Flemish government and with some preferred target groups, the FEN strategy also envisaged communication with and participation of 'civil society', in particular local authorities, target groups and civilians in general. The organization of this 'dialogue' would be outsourced to consultancy agencies. The initial assignment was highly ambitious: 'drawing up, organising and implementing the provision of information to and the deliberation with several target groups during the designation procedure of FEN 1st stage, including the production of materials and documents'. The original assignment was twofold: part of the communication would be directed at several societal groups during the FEN's first stage of designation, and another part would consist of communication training for government officials from the various governmental bodies involved. We will only discuss the first part. This part implied much more than mere information provision. Organised unilaterally from government to society, the assignment aimed at what was called a 'real consultation of society'. Moreover, the government explicitly planned unanticipated forms of communication and deliberation with several target groups (policy makers, people responsible for the implementation, landowners, users etc.).

14.4.2 The Policy Process: Pro-active Participation Reduced to Public Enquiry

The policy process as it was actually carried out differed in several ways from that initially envisaged, especially as far as the goals on participation were concerned. From a strategic perspective, the lack of experience in participation, on one hand, and the constraining political circumstances, on the other, were the main reasons for these differences. From an institutional perspective, though, we will point below at the incompatibility of two competing patterns of (organising) public participation.

As stated above, the Nature Decree initially foresaw two different stages with two different processes that would delimit the FEN while each process depended upon the land use as already acknowledged by spatial planning. For the areas that were already 'green', the process was restricted to an advisory round within the Flemish administration. From a nature conservation perspective, this probably sounds like a light procedure, which raises the question why it was subsequently enhanced with an ambitious and ambiguous participation plan. In addition, it was an unanticipated ad hoc procedure, specifically designed for this particular part of nature policy. Regardless of the reasons for doing it this way, it differed from what is common in spatial planning. Indeed, spatial planning normally entails a public enquiry procedure, which was now applied to the delimitation of the FEN in areas with 'other than green' land uses. In brief, part of the procedure envisaged with regard to the FEN, already ambiguous on 'light' and 'enhanced', was in conflict, and in fact incompatible, with regular spatial planning procedures. Unsurprisingly, this was taken up by those opposed to the FEN idea, either in general or with regard to the land they owned or used.

Therefore, the preliminary decision reached by the Flemish government on 19 July 2002 about the designation of the FEN was accompanied by an amendment of the Nature Decree, in the sense that the so-called special protection zones, based upon the European Bird and Habitat Directive, were added to the decree. As to the process, the designation procedure for FEN and IVON was thoroughly modified; a public enquiry was (still) required regardless of the actual use of the land. In other words, the exceptional participation procedure for 'already green areas' was put aside and replaced by regular spatial planning procedures.

As is obvious, this modification was the (temporary) result of a (still) ongoing political battle. Although after the disaster of the GMS the protests against a network of green areas also slowed down, a new initiative in this field was not popular. Within the Flemish government (1999–2003), the green party, Agalev, and its minister for environment, nature and agriculture, Ms. Dua, confronted the liberal VLD and its minister for spatial planning, Mr. Van Mechelen, who was backed by his fellow party member and then minister-president Dewael. Within the cabinet, the FEN was a highly controversial topic as is clear from the fact that the FEN was included in the so-called Summer Agreement of 2002, a sort of political adjustment of the 1999 Government Agreement. In this Summer Agreement, the majority parties confirmed the allocation objectives as formulated in the Nature Decree and the Spatial Structure Plan for Flanders, but on the condition that a public enquiry be planned for the whole area, including the 'already green' ones, thus reopening the societal and political debate on their destination and future. This decision thus harked back to the traditional rules for allocation and location of different forms of land use as determined in spatial planning: a public enquiry about land use, from which nature was not excluded. Simultaneously, the originally planned request for recommendations from local authorities was cancelled, as was the communication with and participation of the general public. The latter was clearly incompatible with the initial ambitions on participation.

Paradoxically, however, this decrease in participation ambitions increased the opportunities for all kinds of participants to obstruct the process. That was exactly what the liberal party VLD and other opponents of the FEN had wanted from the outset. In our view, the opponents of the FEN seem to have a particular strong argument in this case: why should nature claims be treated differently from other spatial claims? It appears that the people who designed the exceptional procedure assumed that, once the allocation of land use ('green') was settled – and they assumed it was – the remaining issues could be solved by a process of information and deliberation in the field. They also thought that this would be beneficial to the support and legitimacy of nature policy. For the opponents of the FEN, the land use debate, however, even in so-called 'green areas', was far from settled. They wanted to open up the participation process in such a fashion that not only the designation of green areas, but also the principles of FEN itself, including its allocation of hectares, could be debated. One can imagine the frustration of the consultants, civil servants and others who had worked so hard to set up a participation strategy, which now had to be abandoned.

In brief, despite the lessons drawn from the GMS case, once again the agents involved in nature policy making underwent an unhappy experience with processes of advice, consultation and participation. Again, it was from a lack of expertise with participation processes, in this case under constraining political circumstances. But, it was also from the inadequate handling of two adjacent but in some respects competing policy domains.

14.4.3 The Final Outcome: A Public Enquiry

From 23 September to 21 November 2002, the areas initially nominated to be part of the FEN were presented to the population in a public enquiry. Simultaneously (September–October 2002), the Flemish government launched its communication campaign, which was now reduced to a one-way transfer of information. On the one hand, the campaign consisted of distributing flyers (112,400 copies), a general brochure (about 32,000 copies), and brochures for different target groups or stakeholders (agriculture: 13,400 copies; landowners: 13,400 copies; recreation sector: 15,200 copies; and forestry: 12,200 copies). On the other hand, provincial FEN information markets (about 900 visitors) were organised and a website was launched (about 11,000 hits). The government regarded the information markets, initially planned as exchange of information within a larger participation strategy, as threatening and thus invested little efforts in them. Finally, some information days were organised for government officials and local authorities.

A total of 8,968 objections to the FEN plans were submitted, 1,032 of which nonetheless supported the new nature policy. Furthermore, some 3,000 objections were identical and 458 came from governmental bodies. Finally, a number of objections were submitted under a false or fictitious name. It is a bit far-fetched to speak of 'sabotage', as the Flemish newspaper De Morgen (17.05.2003) did, but the number of objections clearly proved that the opposition against the FEN was organised.

From the objections we draw a number of observations. Regarding their content, one has to conclude that the objections addressed practically everything. This raises the fundamental question whether a public enquiry about something such as complex as the FEN is suitable in the first place. The objections also raise the practical and political question whether the way in which it was framed (what exactly was submitted to public enquiry?) would not have required more precision and specificity, or whether it was deliberately kept as broad as it was. There are also observations to be made about the process. First, a large amount of objections came from the government itself, usually from institutes located in adjacent policy domains. This reflects insufficient coordination within the state apparatus itself. Second, many objections were lodged by (groups of) actors that had not been involved in the earlier stages of the FEN. As mentioned above, the process did not allow for broad deliberation. Input was restricted to a small number of agents. Other affected parties, among which the landowners, had not been involved at all. Thirdly, because there were no other available means to make themselves heard, provinces and local communities too felt compelled to clarify their points through formal objection, often with the subsidiarity principle as an argument. In other words, precisely those agents that were excluded from the process, tried to reinforce their position by means of the public enquiry, which thereby canalised political frustrations over earlier lack of access.

The nature of the objections, as well as the actors lodging them, reflect some of the remarkable characteristics of the process itself. As already said, the objections originating from the governmental agencies and local authorities demonstrate a lack of formal, vertical and horizontal coordination within the state apparatus; a public investigation appears to be the forum for deliberation and collaboration with agents from other policy domains and other levels of government. This is certainly true for formal planning procedures, as in spatial planning. Apart from some other shortcomings already commented upon, the FEN procedure also did not anticipate the parallel processes of the 'structure planning' – a renewed set of practices in local spatial planning – with local authorities. At the same time, preliminary deliberation with other than well known target groups was not foreseen, although similar linkages were initiated in other large land use decision making processes (harbour expansion, large-scale infrastructure and others).

Once more, these features and issues indicate the hybrid character of the process: at first, specific and exceptional rules were made up for the first stage of the FEN, with the general rulings of spatial planning to be applied to only a part of it. When this did not met with sufficient political support, the generic rules of spatial planning were applied instead, however in an impromptu way. As a result, there were incompatible promises and practices of participation, which were to be solved in one go in an (unclear) process that was meant to address difficult issues of the allocation, the location and the regime of areas designated for 'nature' at the Flemish as well as at the local level. Hence, it should not come as a surprise that enquiry faced severe criticism, even from people supporting the FEN. Moreover, it caused a great deal of harm to the overall legitimacy of Flemish nature policy. As was the case for the GMS, the maps used for the FEN were prematurely distributed, causing great societal unrest. Furthermore, the limited distribution of the FEN maps, contrasting to the large distribution of the maps of the 'Desired Agrarian Structure' (*'Gewenste Agrarische Structuur', GAS*), furthered the distrust among different stakeholders from Flemish rural areas.

Reflecting the societal controversies surrounding the FEN, 20,000 people, organized under the label of the 'Countryside Platform', gathered on Sunday 11 May, 2003 in Ghent for a huge demonstration against the FEN. Together with the local demonstrations, the Ghent one clearly expressed dissatisfaction about local designation issues. While partly motivated by NIMBY-preferences, the demonstrations also expressed dissatisfaction about lack of openness, lack of information, and lack of participation possibilities as well as fears about restrictive land usages. Several agents were afraid that joint usages would be prohibited overall, so that the FEN actually implied the construction of 'pure' nature. In other words, political contradictions and ad hoc changes, both of content and process, as well as shifts in goals, incompatible deadlines, unannounced switches from one mode of participation to another, severely affected the societal legitimacy of the FEN. At the end of the day, because of sloppy organising it was unclear which issue or question would be decided at which stage, by whom and at which level, and with what procedure. The resulting lack of legitimacy is indicated by the pattern of social mobilisation, referred to above, as well as the press analysis presented below.

14.4.4 A Newspaper Analysis: The Shifting Coverage
of Shifting Arguments

In order to gain insight in the way the FEN was presented to the general public, we carried out a content analysis of press clippings on the FEN in the Flemish newspapers from May 2001 to May 2003 (De Zitter et al. 2003). Based on over 300 articles, this analysis gives an outline of the shifting news coverage of the shifting arguments surrounding the FEN.

First, a quantitative analysis of press reports shows three peaks in news coverage that parallel the political and societal debate: June 2002 (preliminary decision-making within the Flemish government, Summer Agreement), November 2002 (end of the public enquiry) and May 2003 (demonstration in Ghent, federal elections).

Second, during the period 2001–2003, a clear shift occured in the focus of the reporting. From general and relatively positive articles at the beginning, the reporting gradually started to focus upon the problems in specific locations. Newspapers with regional editions or pages led the way. The steady accumulation of these local issues clearly indicates a gradual escalation of the discussion; from regional pages and issues, the reports gradually shifted to larger political contestation over the overall spatial claims for nature at the Flemish level as well as towards the way decisions were made. The public enquiry functioned as catalyst. For it was during the enquiry that it became clear to the parties involved whether or not their property, area or territory would become part of the FEN. The (unclear) maps that were used led to an increased sense of unrest and mobilisation.

This local unrest was enhanced by conflicts within the Flemish government. Statements by the minister–president (Dewael, VLD) and the minister for spatial planning (Van Mechelen, VLD) were contradicted by those of the minister for the environment (Dua, then Agalev, now Groen!). The dissensus was not only about substantial FEN targets, but also about the way decisions were made. One government party (SP.a) distantiated itself from this debate, as did the opposition parties (CD&V, Vlaams Blok). However, the latter were present at the demonstration in Ghent, in May 2003, one week before the federal elections.

Content analysis also demonstrates the dominance of certain actors and their viewpoints. In particular, fishermen, hunters, forest owners and farmers were the ones getting their voices heard most easily. The accumulation of objections formulated against the FEN gradually led to an ad hoc coalition of these interest groups. Eventually, from location and joint use issues, the coalition finally succeeded in reopening the discussion about the larger goals, surfaces and protection regimes of the FEN. Advocates of the FEN were hardly present in the newspapers. There was no public support from the majority political parties, while the nature conservation movement and the scientific world also failed to express their support publicly. This could be because editorial boards simply refused to print their viewpoints. This is contradicted, however, by the observation that the nature movement initially did succeed in gaining public recognition for the general goals of the FEN and subsequently was able to voice complaints about the lack of progress. Along the way, the movement's attention turned to the growing insecurity of other actors involved. The

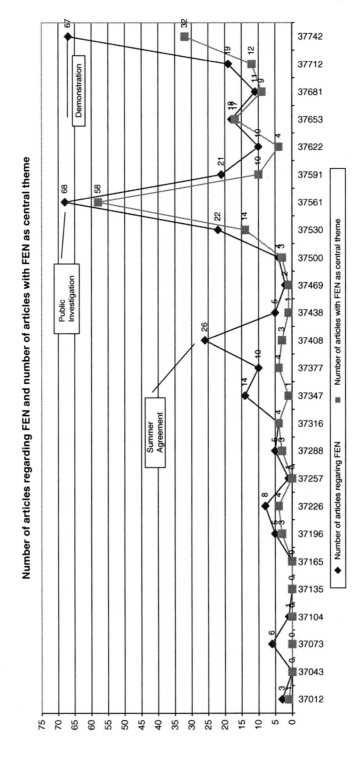

Number of articles regarding FEN and number of articles with FEN as central theme

issue of joint use even led to press articles in which the nature movement publicly supported the fishermen.

14.4.5 The Eventual Outcome: A Demonstration and New Political Promises

Content analysis confirms that, at a certain moment in time, the only person actually defending the FEN was the responsible minister. The public protest culminated in the demonstration of 11 May 2003, organised by a coalition that went by the name 'Countryside Platform', which objected primarily to what it called the 'patronization' and the 'over-regularisation' of the countryside. The date of 11 May was picked for good reasons, since federal parliamentary elections were planned for the end of the month. The FEN was and still is a Flemish matter, and, therefore, outside the reach of the federal level. However, the Ghent demonstration did without doubt strongly influence the federal election result of the Green Party *Agalev*. During the elections, the Green Party lost all its seats in parliament! Under the new name *Groen!* it barely survived in the Flemish elections since then.

Concerning the decision-making and the implementation of the FEN itself, three moments are worth mentioning. First, in March 2003 the MiNa-Council, the regional advisory body for environment and nature, gave advice on the handling of the public enquiry results and the ensuing designation plans. Second, on 27 June 2003 the Flemish government, with a clipped 'green wing' as a result of the federal elections of May that year, approved the draft plan of the FEN. On 18 July 2003, the operational plans were approved and published in the official Belgian Journal 'Staatsblad' on 17 October. This finalised the first stage of the designation process, which was limited to the 'green areas' and the 'consensus areas'. Third, in 2004 the Flemish government confirmed the previously formulated objectives of the Nature Decree (1997) and the SPF (1997).

Still, the implementation of the FEN is and will be problematic. To give just one example of the difficulties ahead: the Nature Decree (1997) determines that a so-called Nature Guideline Plan (NGP, *natuurrichtplan*) should be made for every area of the FEN before 2008. This plan aims at protecting and developing area-specific nature values. It took untill 2005 before the first NGPs were officially announced (Bogaert et al. 2005). A NGP may add some protective measures to a FEN area, which implies new discussions about precise location and joint use agreements. To make up these plans, planning group have been installed by the Nature Department in collaboration with steering groups of local governments and interested parties, such as representatives of nature associations, farmers, forest owners, hunters, fishermen, etc. These NGPs are a crucial tool for the further elaboration, implementation and realisation of nature policy making. The deliberative infrastructure constructed around it, however, makes once again a huge call on the participatory capacities of the agents involved in nature policy. There are, in fact, not only many actors involved in such a steering group, but the decree also foresees a public enquiry for each plan. All these efforts may give nature policy making the chance to clean its tarnished participatory reputation, but they surely do place high demands on the resources and participatory capacities of the agents involved.

14.5 Conclusion

In political and social sciences literature, legitimacy has been defined, operationalised and measured in various ways (compare, among many others, Cashore 2002; Parkinson 2003; Aarts 2004). In this contribution, the concept of legitimacy was applied to specific policies, formulated in a specific polity and related to a specific policy domain, i.e. nature policy making. Secondly, the concept was operationalised in terms of the extent to which the parties involved perceive the policy as morally authoritative and acceptable, possessing a clear and correct content, and as being embedded in a uniform and manageable implementation process. A particular policy, thus, is legitimate if it meets these conditions.

On the basis of this definition, one cannot but conclude that Flemish nature policy over the last decade or so has been subject to a steady erosion of legitimacy. We believe, though, that it is not so much its content that causes the problem. Of course, people and societal organisations differ and will continue to differ with regard to their preferences over the qualities of the environment, the future of rural areas, and they surely differ on how to weigh agriculture, nature, hunting and other activities. Nonetheless, we believe that it is not this eternal controversy over content that has caused legitimacy decline in Flanders, since in the end a fair and pragmatic settlement between competing values and goals can be accepted by different parties. Rather, it is the lack of clarity and transparency that provokes irritation and unrest. On top of that, recent Flemish nature policy has handled and managed nature policy processes in a rather awkward way, in particular regarding the compatibility of GMS and FEN. Our analysis provides ample illustrations of the sloppy management of the launching, withdrawal, amending and redesigning of strategies, even with regard to basic questions such as: how to proceed, how to decide who will be involved, when and at what level what decision on which issue will be taken? The inconsistent answers given to these and related questions have resulted in a situation in which nature policy planning is perceived by the public as unpredictable and unreliable.

Of course, the debates, procedures and decision making on both GMS and FEN have been affected by circumstantial factors: the coincidental linkages with unrest in the adjacent field of agriculture, e.g. manure policies, as well as the contingent confluence of the FEN with a renewal in the spatial planning process. It also seems fair to add old-fashioned controversies and harassments between differing political parties to these circumstantial factors. However, while these factors might be circumstantial they are by no means incidental. To the contrary. We believe that they are the outgrowth of two basic weaknesses of current Flemish nature policy making. First, its inadequacy in dealing with questions of how to accommodate conflicting claims originating from adjacent policy fields. Second, its inexperience with participatory decision making and its requirements in terms of inclusiveness, resources and capacities. Current Flemish nature policies lack a clear vision, a sustained strategy, an organisational infrastructure and the expertise for an adequate handling of these issues. Like other environmental policy domains, the more nature policy develops into a multi-actor, multi-sector and multi-layered game, the more these issues will be at the forefront of policy making, turning the search for new abilities of the political personnel staffing these policy making processes into an urgent issue.

References

Aarts, N. (2004). 'Communicatiewetenschappelijke kijk op verantwoordelijkheid voor natuur'. In G. Overbeek en S. Lijmbach, red. *Medeverantwoordelijkheid voor natuur*. Wageningen: Wageningen Academic Publishers, pp. 71–85.

Blust, G. de, D. Paelinckx and E. Kuijken (1995). 'The Green Main Structure For Flanders. The development and implementation of an ecological network'. *Landschap* 12: 89–98.

Bogaert, D. (2004). *Natuurbeleid in Vlaanderen. Natuurontwikkeling en draagvlak als vernieuwingen?*. Nijmegen/Brussels: Instituut voor Natuurbehoud.

Bogaert, D., and J. Gersie (2006). 'Recent nature policy dynamics in the Netherlands and in Flanders.' In B. Arts and P. Leroy eds. *Institutional Dynamics in Environmental Governance*. Dordrecht: Springer, pp. 115–138.

Bogaert, D., J. Lambrecht, A. Cliquet and G. Van Hoorick (2005). 'Gebiedsgerichte plannen. Natuurrichtplannen.' In A. Cliquet, G. van Hoorick, J. Lambrecht and D. Bogaert. *Gebiedsgericht natuurbeleid: operationalisering en uitvoering van de Vogelrichtlijn en de Habitatrichtlijn*, in MIRA-BE 2005, Milieurapport Vlaanderen: Beleidsevaluatie, VMM, pp. 98–102.

Cashore, B. (2002). 'Legitimacy and the privatization of environmental governance.' *Governance* 15(4): 503–529.

Celen, G., L. Daniëls and J. Verheeke (1990). 'Naar een natuurbeleid voor de jaren negentig.' *Leefmilieu* 1: 17–23.

Cliquet, A., G. van Hoorick, J. Lambrecht and D. Bogaert (2005). *Gebiedsgericht natuurbeleid: operationalisering en uitvoering van de Vogelrichtlijn en de Habitatrichtlijn*, in MIRA-BE 2005, Milieurapport Vlaanderen: Beleidsevaluatie, VMM, pp. 81–110.

Hoorick, G. van (2000). *Juridische aspecten van het natuurbehoud en de landschapszorg*. Antwerpen-Groningen: Intersentia.

Jong, D. de (1999). *Tussen natuurontwikkeling en Landschaftsschutz. Sociaal-cognitieve configuraties in het grensoverschrijdende natuurbeleid*. Nijmegen/Den Haag: Eburon

Kersbergen, K. van, and F. van Waarden (2001). *Shifts in governance: Problems of Legitimacy and Accountability*. Den Haag: NWO.

Kuijken, E. (1988). 'De plaats van natuurbehoud in het ruimtelijk en milieubeleid.' In *Ruimte voor Groen. Deel I, vijfde Vlaams wetenschappelijk Congres over Groenvoorziening*. Gent: Vereniging voor groenvoorziening vzw, pp. 129–146.

Kuijken, E. (1996). 'Vragen rond natuurbehoud: achtergronden en recente beleidskaders.' In *Hof van Eden of Toren van Babel? Natuurbehoud en Natuurontwikkeling in Vlaanderen*. KBS, pp. 6–23.

Leroy, P., and D. Bogaert (2004) 'Het Vlaams Ecologisch Netwerk als natuurconflict: Verwarde beleidsvoering en georganiseerde hindermacht.' *Landschap* 21(4): 211–224.

Leroy, T., I. Loots and P. Leroy (2004). *Natuur: hoe kan dat? Onderzoek naar het maatschappelijk draagvlak voor natuurbehoud in Vlaanderen*. Antwerpen: Universiteit Antwerpen.

Marks, G., L. Hooghe and K. Blank (1996). 'European integration form the 1980's: state-centric v. multi-level governance.' *Journal of Common Market Studies* 34(3): 341–78.

Parkinson, J. (2003). 'Legitimacy problems in deliberative democracy.' *Political Studies* 51(1): 180–196.

Pierre, J., ed. (2000). *Debating Governance. Authority, Steering, and Democracy.* Oxford: Oxford University Press.

Pinton, F. (ed.), P. Alphandery, J-P. Billaud, C. Deverre, A. Fortier and G. Géniaux (2005). *Scènes locales de concertation autour de la nature. La construction Française du réseau Natura 2000.* Paris: Document Final, IFB, MEDD.

Verheeke, J. (1990). 'Een nieuw decennium natuurbehoud. Naar een "model Japan" of naar een "offensief natuurbeleid".' *Leefmilieu* 5: 133–138.

Wiering, M. (1999). *Controleurs in context – Handhaving van mestwetgeving in Nederland en Vlaanderen.* Lelystad: Koninklijke Vermande.

Windt, H. J. van der (1995). *En dan: wat is natuur nog in dit land? Natuurbescherming in Nederland 1880–1990.* Amsterdam/Meppel: Boom.

Zitter, M. de, T. Wymeersch and D. Bogaert (2003). 'Publiek draagvlak voor het Vlaams Ecologisch Netwerk: een kwantitatieve én kwalitatieve benadering van de VEN-berichtgeving in de Vlaamse dagbladpers.' In M. De Zitter, T. Wymeersch, D. Bogaert and A. Cliquet. *Onderzoek naar het draagvlak voor natuurbehoud in Vlaanderen. Deelopdracht: Opstellen van een korf van indicatoren voor het meten van het maatschappelijk draagvlak voor natuurbehoud in Vlaanderen.* Ghent: Arteveldehogeschool/University of Ghent, pp. 123–188.

Zouwen, M. van der and J.P.M. van Tatenhove (2002). *Implementatie van Europees natuurbeleid in Nederland.* Wageningen: Natuurplanbureau.

Chapter 15
Nature Policy in Flanders

Messy Participation Strategies or Unfavourable Circumstances?

Erik van Zadelhoff

Bogaert and Leroy judge Flanders Nature Policy as messy. According to them, a clear vision, a sustained strategy and the expertise to handle complex policy processes is missing. A thorough course in multi-actor, multi-level and multi-layer policy-making for policy personnel should be the cure for this. From my own experience in nature policy in the Netherlands, and with some understanding of the Flemish situation, I strongly doubt that such a course would be very helpful. There is more to the poor success of nature policy in Flanders than just some shortcomings in the policy process.

To explain this, a comparison of the situation in the Netherlands and Flanders is useful. To begin with, there are striking similarities between both countries. Both have a very high population density, an intensive land use and only scattered pieces of land that can still be considered as 'natural'. Nature is scarce and vulnerable in both countries, and people are aware of that. Over the last decades 'the environment' has received a great deal of attention in both countries. There is substantial support for protection of the environment, both in Flanders and the Netherlands. This is reflected in the support for Green Parties and the way parliamentarians vote in the European Parliament. Belgian green parties (Groen and Ecolo) gained over 10% of votes for the national parliament in 2003; in the Netherlands this was slightly more than 5% (www.europeangreens.org). In terms of eco-friendly voting, the Netherlands and Belgium ranked 3rd and 4th respectively among the 15 'old' member states of the EU (www.foeeurope.org).

It is no surprise that, under these circumstances, both countries developed similar solutions to combat the process of deterioration of nature. Conservationists in both countries realized that the only really effective strategy to prevent the isolation of small nature reserves and to stop the lowering of water tables, the influx of nutrients by (ground) water etcetera, is to realise a robust ecological network. At the European

Erik van Zadelhoff
IUCN, World Conservation Union, Brussels and Amsterdam
E-mail: f.vanzadelhoff@chello.nl

J. Keulartz and G. Leistra (eds.), *Legitimacy in European Nature Conservation Policy:*
Case Studies in Multilevel Governance, 205–207. © Springer 2008

level this strategy is also accepted and supported (Natura 2000). From this point of view, one cannot say that the Flanders Nature Policy lacks vision, or that it does not have a sustained strategy. On the contrary, the very fact that the Flemish Ecological Network was initiated shortly after the failure of the Green Main Structure is a clear example of a sustained strategy. And, from its content, it is evident that this strategy was chosen for good reasons. The really interesting question to ask is, why did the same strategy fail in Flanders when it was, in general terms, a success in the Netherlands.[1] All the more so, since the policy process in the Netherlands was certainly not more participative than it was in Flanders.

Three main differences between the Netherlands and Flanders can explain this.

Firstly, there is a marked difference in public support for nature conservation in the strict sense. In the Netherlands, nature conservation started to be a public issue from the early 1900s on. Natuurmonumenten was established in 1905, and the first nature reserve, the Naardermeer, was acquired in 1906. From the beginning, public campaigning comprised an important part of the activities, resulting in over 900,000 paying members today. In Flanders, there was hardly any organized public nature conservation until the mid 1900s. At the present time, about 1% of the Flemish population are paying members of nature reserves managing NGOs, whilst in the Netherlands this is about 10%. The same difference can be seen for the total area of protected nature; some 2% in Flanders compared to over 10% in the Netherlands.[2] Although the large NGOs in the Netherlands were, like their Flemish counterparts, not actively advocating the realisation of the Ecological Network in the media, they had a marked influence behind the scenes on policy makers. Moreover, the concept was advocated to their members through internal membership communication, resulting in silent support from a substantial proportion of the (urban) public.

Secondly, the Netherlands has a much stronger tradition in spatial planning than Flanders. In the Netherlands, well-elaborated procedures and strict rules exist, not only with regard to urban planning, but also concerning rural areas. These procedures are well developed at all levels; local, provincial and national. Therefore, nature policy in the Netherlands was embedded in a dominant system of spatial planning and a relatively strong tradition of large scale planning on the part of the central government, whilst in Flanders this was much less the case. Because of this dominant system, the perception in the Netherlands at the time the National Ecological Network was announced officially was that there was neither the need, nor the space, for participative decision-making experiments. The tradition of spatial planning was therefore an important source of legitimacy for the Ecological Network in the Netherlands, a source that was largely lacking in Flanders.

A third important difference between the two countries has been the political context over the last decades. As Bogaert and Leroy explain, in Flanders a Minister for Environment from a green party defended the Ecological Network. This Minister was in an isolated position because of lack of support from her colleagues among

---

[1] Milieu en Natuurplanbureau, Natuurbalans 2004 en 2005. RIVM Bilthoven.

[2] www.natuurmonumenten.nl; www.landschappen.nl; Staatsbosbeheer, Jaarverslag, 2004.

the Christian Democrats and Liberals in the Flemish Government. This process of isolation occurred against the background of the need for more strict rules on the use of manure on arable land, imposed by the European Commission.

In the Netherlands, the National Ecological Network was launched in 1990 in the first Nature Policy Plan. The Government at the time was of a Christian Democratic/Socialist character, with a Christian Democratic Minister of Agriculture, Nature Management and Fisheries (LNV). This minister was also confronted with the need for stricter rules on the use of manure. The difference in this case was, however, that this Minister had to impose strict measures on the traditional clientele of his party, the farmers and fishermen. Under the influence of the Prime Minister from the same party, the choice was made to withstand the pressure from this traditional clientele and to take real responsibility for the protection of the environment. This was considered to be in the long-term interests of the farmers and an expression of good (Christian) stewardship for nature. This course received the support of the large majority of parliament. The green party only played a marginal role in the debate. Unlike in the Flemish situation, the Christian Democrats, as the main political power, decided to follow a pro-environment course. By doing so, the environment was no longer the domain of left wing one-issue parties, but became a mainstream political topic.

Taking these three differences into account, one can hardly be surprised that the concept of realizing an ecological network in Flanders failed. Being a logical concept and strategy from an ecological point of view, the context was not there just to obtain enough political and public support. To think that a concept that proved to be successful in one place could be implemented in another place with a quite different social context was a serious misjudgement, rather than a case of mismanagement. The real challenge for Flanders is to solve the very great difference in support for the environment in Flemish society at the general level and the lack of support for practical measures at the national, regional and local level. A challenge not just for policy makers, but also for civil society as a whole.

Puisaye Ponds in Burgundy (Cedric Foutel)

Chapter 16
Between European Injunction and Local Consultation

Analysing the Territorialization Process for a Public Nature Conservation Initiative in France[1]

Florence Pinton

16.1 Introduction

After a long period of favouring areas free of all human activity as a means of protecting nature areas, nature conservationists have begun to question this approach in the face of new paradigms within environmental science, combined with relatively inconclusive forms of action in the field. By acknowledging the central role played by local populations in the management of natural resources, the characterization and understanding of local knowledge have gradually become key objectives of any conservation programme, and are now considered to be essential to the success of participative management projects. Especially in Southern countries, where nature only manifests itself in small strips due to the extent of human intervention, the importance of local practices and socio-economic constraints have increasingly become recognized by local leaders. In France, environmental policies have progressed hand-in-hand with the so-called 'environmentalization' of vast areas of rural land previously used mainly for production activities, a process that Hervieu and Viard (1996) have qualified as 'publicization' of the countryside. In this process the

Florence Pinton
Department of Sociology, University of Nanterre (Paris 10)
E-mail: florence.pinton@orleans.ird.fr

[1] This chapter is based on the results of research carried out by the GRENAT network – *Groupe de Recherches Sociologiques sur la Nature* (Sociological Nature Research Group) – coordinated by the author (Pinton et al. 2007). The researchers involved in GRENAT (P. Alphandéry, J.-P. Billaud, C. Deverre, A. Fortier, G. Géniaux, N. Perrot and F. Pinton) are sociologists who have been involved in monitoring the application of the Habitats Directive in France for over 10 years, and who focus on the evolution of and the expertise mobilized for nature policies. Their project has been jointly funded by the French Ministry of the Environment and Sustainable Development (MEDD) and the Institut Français de la Biodiversité (IFB) (French Institute of Biodiversity).

right to inspect and administer rural areas is increasingly transferred from farmers and landowners to environmental agencies.

Adopted in 1992 by the EU, the aim of the Habitats Directive is to preserve European natural heritage by developing an ecological network, Natura 2000. In order to reach the objectives set out in the Habitats Directive, the EU adopted a schedule organized into three stages. In the first stage, each member state was to propose a list of sites, that had to be submitted to the Commission by June 1995. In the second stage, the Commission would establish a draft list of Sites of Community Importance (SCI). The third stage was aimed at the official designation of these sites by each member state and their classification into Special Areas of Conservation (SAC). Part of this final stage was the definition and development of management plans for the protection of these sites. For this final stage, France has chosen a contractual approach based on so-called 'documents of objectives' Once drafted, these documents of objectives had to be approved by decree and then contracts had to be signed by the environmental and socio-economic stakeholders (farmers, foresters, fishermen) involved.[2]

Application of these documents has caused much controversy. More specifically, the scientific knowledge that was mobilized and its use in the field have generated considerable disagreement between stakeholders, leading to a scientific and political crisis. To resume dialogue with the different stakeholders after a period of severe conflict between 1995 and 1997, the Ministry of the Environment thoroughly revised the procedure. After several years of experimentation, the French system meanwhile appears to be locally accepted, but the number of areas selected for the ecological network has decreased significantly compared to the number anticipated by scientists.

The aim of this chapter is to assess the French process of proposing sites, with a focus on the creation of documents of objectives in the third and final stage of the process. The European dimension of the directive makes biodiversity a tricky injunction to implement at a local level. How can a European system of reference for nature conservation tie in with local cultures? And how can a spatial policy that defines objectives at a European level, be translated into a regional policy based on the local use of nature? This article seeks to answer such questions based on French experience.

16.2 Redefining the Framework for Public Action

16.2.1 From Injunction to Contention

The difficulty in applying the Habitats Directive partly lies in the paradox created by demanding that it should be based on science only but at the same time claiming

---

[2] France and the UK are the only EU member states that have developed contractual approaches for all activities taking place within the sites.

that it should also take account of social and economic issues. A sharp distinction was made between the compilation of national inventories by scientists and their validation at the European level in the first two stages on the one hand, and the consultation involving other rural stakeholders in defining, drawing and developing management plans for these sites in the last stage on the other hand. To use the words of Dodier, this distinction reflects a desire to isolate environmental institutions. It was 'a process that consists of sheltering the interior of an institution from the pressures emanating from its exterior' (Dodier 2003: 61). From a political point of view, the inventory procedure adopted by France was planned as if it was to take place in a stable world free from scientific and social disagreement (Godard 2002).

In fact, this inventory procedure provoked a lot of resistance, particularly from 1995 to 1997, causing a scientific and political crisis that called into question the very institutional framework proposed by the Ministry of the Environment. Opposition reached its peak in March 1996, when the administration presented its final list, including 1,316 sites covering a surface of 7 million hectares, i.e. 13% of the country. The opposition movement, the so-called 'group of nine', a powerful alliance of hunters, foresters, farmers and fishermen, fiercely objected to this list. Their action led to a suspension of the inventory procedure for several months in 1997.

The group of nine contested the scientific credibility and reliability of the inventory procedure (Alphandéry and Fortier 2001). In fact, the scientific basis of the procedure was uncertain and incomplete. 'The people who signed the documents overestimated the state of knowledge in France', we were told during our first surveys (Rémy et al. 1999). In the field, the lack of reliable data led operators to rely on the inventory of ZNIEFF (Natural Areas of Ecological and Wildlife Interest), which was established in 1982 by the Muséum National d'Histoire Naturelle (French Natural History Museum) with the aim of carrying out a preliminary inventory of French natural heritage. Resorting to this source as the single point of departure was a bone of contention for local stakeholders, who condemned its lack of scientific rigour. It also revived memories of the numerous disagreements provoked by the past use of these inventories by nature conservation NGOs. Representatives of the local stakeholders criticized both the methods and the conclusions of this scientific assessment and have demanded a second assessment in which they themselves would be involved.

16.2.2 Opening up the Procedure to Stakeholders

The rural protest movement brought about a thorough modification of the procedure, with the opening of a forum for negotiation, the principal result of which was a possible revision of the site boundaries in conjunction with 'non-scientific' stakeholders concerned with the Natura 2000 project. Political measures were adopted to take into account different kinds of knowledge in order to redefine the sites and their boundaries in a consensual manner. The end result would appear to be an increase in institutional openness. The time invested in the process, which was much longer than that originally planned by the EU, has allowed for the observations and criticisms made by non-scientists, who wanted to have a say at every stage of the procedure and not just at the final stage (the implementation of management measures).

To renegotiate the sites and their perimeters, it was necessary to transgress the boundary between the sphere of science and that of action and to link the production of scientific tools to a deliberative framework. As result of the consultations with non-scientific stakeholders, the total area of sites proposed to the EU dropped from the initial figure of 13% of the country in 1996 to 7.6% at the end of 2002.[3] France was being stigmatized by European observers for the delay involved in the submission of sites and the relatively small percentage of areas proposed, compared to the percentages of areas proposed by other countries. However, such percentages only have limited meaning.

By solely referring to the size of the area designated by each country under the Habitats Directive or the Birds Directive in order to identify complying countries (Table 16.1), the European Commission groups together data that are not necessarily of equal significance. In our opinion, such percentages remain insufficient and incomplete. They say nothing about the expansion of protected areas in relation to the initial situation, the social repercussions of zoning, or the scope and effectiveness of the management measures to be applied. In other words, the percentages do not reveal the real social and environmental consequences of the implementation of Natura 2000. If we use other indicators as a basis for our reasoning, such as the number of communes and inhabitants directly or indirectly affected by the process, the sheer size of sites alone is less relevant. On the national scale, over a third of French communes have a Natura 2000 site, or part of one, within their boundaries. So there is a relatively high social impact involved, in the sense that the local councillors of these communes are obliged to take part in the steering committees supervising the creation of documents of objectives (Pinton et al. 2007: 67).

Table 16.1 The Natura 2000 barometer on the Habitats Directive of 28 March 2003 (European Commission Newsletter, DG ENV, n° 16)

Member State (not exhaustive)	Number of sites proposed	Total area proposed (km^2)	% of national territory
France	1 202	41 295	7.6
United Kingdom	567	24 064	9.9
Belgium	270	3 178	10.4
Sweden	3420	57 476	12.8
Italy	2369	41 261	13.7
Luxembourg	38	352	13.7
Portugal	94	16 500	17.9
Spain	1276	118 496	23.5
Denmark	194	10 259	23.8

---

[3] Besides these 7.6% of Special Areas of Conservation (SACs) submitted under the Habitats Directive, France also designated 2.6% of its territory as Special Protection Areas (SPAs) under the Birds Directive. With a total size of designated areas of approximately 10% of its territory, France is at the bottom of EU member states.

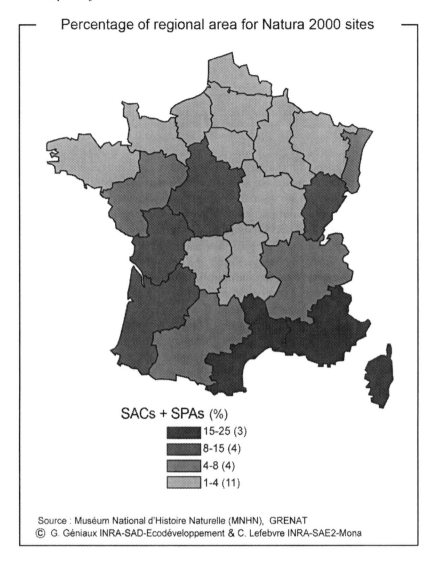

Percentage of regional area for Natura 2000 sites

SACs + SPAs (%)
- 15-25 (3)
- 8-15 (4)
- 4-8 (4)
- 1-4 (11)

Source : Muséum National d'Histoire Naturelle (MNHN), GRENAT
© G. Géniaux INRA-SAD-Ecodéveloppement & C. Lefebvre INRA-SAE2-Mona

16.3 The French Version of the Future Natura 2000 Network

The map of Natura 2000 sites on French territory, produced by GRENAT, reveals that these sites are distributed unequally throughout the country. We also found an unequal site distribution in the four regions in which we did qualitative research. Nord-Pas-de-Calais (NPC) and Burgundy are among the regions with a density of sites below the national average, while Poitou-Charentes (PC) and Provence-Alpes-Côte d'Azur (PACA) have a high coverage of sites (see Table 16.2).

Table 16.2 Percentage of four regional areas

Sites proposed Monitored by:	Burgundy *P. Alphandery and F. Pinton*		NPC *A. Fortier*		PC *J.P. Billaud*		PACA *C. Deverre and N. Perrot*	
	before	after	before	after	before	after	before	after
Number	56	46	32	28	71	53	103	75
% regional area	5	1,74	13	1,47			38,5	19

Our qualitative research carried out in different regions (Pinton et al. 2007) shows that this unequal site distribution not only mirrors local differences in ecological en environmental circumstances but also local differences in power.

The low percentage of sites in the Nord-Pas-de-Calais region and Burgundy can certainly be explained by the character of their natural habitats. Both regions have few natural areas that qualify as 'outstanding', and most of their habitats have been subjected to human intervention and farming. Both regions also have limited experience in the protection of nature, with environmental NGO's that remain weak and poorly linked. However, the final choice of sites in each region was not based on the same strategies or local stakeholders. Fortier has shown that in the Nord-Pas-de-Calais region, the number of sites was dramatically reduced during negotiations, while areas of high value for hunting and game such as wetlands or littoral marshes were even completely removed from the final proposal to the European Commission (Pinton et al. 2007). These changes were mostly the result of a power struggle between the principal administrators of the rural space, who were opposed to Natura 2000, and the institutional actors responsible for applying the directive, i.e. the Regional Directorate for the Environment (DIREN) and the scientists of the Conservatoire Botanique de Bailleul (Bailleul Botanical Conservatory). In a weak position, these last groups were overcome by opponents of the Habitats Directive, who had formed a strong alliance and succeeded in involving their partners in the negotiations thanks to their lobbying capacity.

In Burgundy, on the other hand, the DIREN remained active throughout the entire process with a view to minimizing disputes. Its equally modest result – 1.7% of the regional area – was related to pressure exerted by users of agricultural and forest areas, the influence of hunters and fishermen, and the reticence of the regional council to the application of the Habitats Directive. The power struggles between the different players greatly encouraged the authorities to adopt a cautious approach from the outset and to concentrate on reaching compromises. This choice can also be explained by the inadequacy of regional knowledge concerning nature management. Relevant experience and references were lacking in Burgundy when the Habitats Directive was first implemented. The inventory of natural habitats drawn up within the framework of the Natural Areas of Ecological and Wildlife Interest (ZNIEFFs) proved to be varied and imbalanced according to the department (Alphandéry and Pinton 1999). To remedy this and to establish its legitimacy, the DIREN representative drew up proposals based on a 'consensual scientific' approach characterized by the coexistence of a precise methodological framework and social realism that took into account regional power struggles.

Unlike the Nord-Pas-de-Calais region and Burgundy, the two other regions, Poitou-Charentes (PC) and Provence-Alpes-Côte d'Azur (PACA), were characterized by their territorial ambitions in terms of conservation measures and their experience in negotiating environmental policies. Their regional networks of ecological skills, which were more highly developed and better structured, enabled them to hold strong positions and demonstrate their know-how in emphasizing their point of view.

Our qualitative approach shows that these situations involved contrasting points of view in terms of inventories, associative network, and protected areas. All of them experienced extremely diverse developments at the meeting point of scientific, political, and economic demands and uncertainties. Be that as it may, the biogeographical EU commissions validated all regional proposals despite any disparity or insufficiency in the selections. According to the representative of the Ministry of the Environment the exchange of arguments between scientists and representatives of each of the countries within the commissions were decisive.

The next stage involved producing documents of objectives. This was the stage that should have allowed local users to develop appropriate management measures by means of contractualization. This stage appears as a specific procedure: 'the French response to article 6 of the directive. It was based on 'a decentralized approach to the application of the Habitats Directive' (Valentin-Smith et al. 1998). The roles of administrators and scientists were replaced by those of the socio-professionals and traditional users of the habitats, as well as by local authorities and their representatives. Their involvement contributed to making the directive an important tool for social experimentation in the field of nature policies.

16.4 Documents of Objectives or the Challenge of a New Means of Managing the Rural Space

16.4.1 Associating Cognitive and Deliberative Activities

Testing in the field led French politicians to admit that the two first phases of inventories and delimitation could not be considered in spatial and ecological terms alone, but that they actually lay within the scope of rural areas controlled by historical social groups such as farmers, foresters, and hunters. The decision to implement management objectives on a contractual basis made by the Ministry of the Environment showed the importance that it attached to the local dimension in the definition of the management measures to be established in order to meet environmental objectives.

In this particular case, public action led to the design of a measure – of which the document of objectives was the final result – intended to build a specific relationship between knowledge and action. The stage involving the documents of objectives (DOCOBs) was that of a controlled consultation leading to management proposals that were shared collectively. In a document published by the Ministry of the Environment, the procedure is defined as follows:

> The DOCOB is an innovative procedure. It is drawn up under the responsibility of the departmental prefect assisted by a technical operator, with a significant place given to local consultation. The steering committee, under the authority of the prefect, includes the partners concerned by the management of the site or their representatives. This document defines management directions and contractual conservation measures, and indicates, if need be, the statutory measures to be implemented at the site. It states the means of financing the contractual measures.[4]

The DOCOB involves a top–down strategy that demands results and a bottom–up strategy via consultation. Through the DOCOB, the Ministry of the Environment favours consultation as a tool for action, particularly in stimulating the confrontation of different kinds of approaches related to nature management.

As already stated, no other nature protection policy has ever mobilized such a large number of participants in so many different places. The other side of the coin involves the difficulty related to ensuring the coexistence of extremely diverse knowledge and value registers: biodiversity conservation, on the one hand, local interests and representations, on the other. The whole process demonstrates the will to enact this European policy through the identification of sites at the level of rural territories, i.e., its 'territorialization'.

Investigations carried out by the GRENAT researchers focused on the creation of documents of objectives at the national and local levels and also on the accompanying consultation processes. The aim was to assess the way in which management measures are developed collectively at site level by meeting and confrontation of multiple actors, all of whom held different views on nature and had different knowledge, whether technical, local, or scientific. The first phase of the process (inventories) contrasted scientific ecological knowledge with practical knowledge, whether that of organized groups at the national or regional level, such as hunters, foresters and farmers, or more simply, local knowledge. This second phase, involving the creation of DOCOBs, mobilized different knowledge registers in order to associate them within the definition of management measures. This approach closely links cognitive and deliberative activities, which implies identifying how and on what basis exchanges are established and agreements reached, and beyond that, to question the nature and scope of such agreements. Will the agreements give rise to the production of new knowledge that can be described as 'hybrid' knowledge, and to forms of collective learning capable of modifying the practices of stakeholders?

In a complementary way, the issue of the emergence of a skills network capable of linking different stakeholders and different sites, as well as other European countries, was, in our opinion, a central one. In France, there is a network of protected areas (nature reserves, national parks and 'conservatoires') that could undoubtedly be strengthened by the implementation of the Natura 2000 network. However, it appears inevitable that the network will be forced into competition

---

[4] Natura 2000. *Dix questions dix réponses*, MATE, January 2003, p. 10.

with those organizations traditionally involved in the development of the rural space (farming and forestry organizations, etc.). We can, in fact, ask how different partners will reorganize themselves and to what extent these developments will remodel the management of rural territories in France on a long-term basis.

16.4.2 Integrating Multiple Intervention Scales

To answer the above questions, understand the diversity of situations involved, and solve the problems of scale posed by this kind of investigation, the GRENAT team chose to combine a qualitative approach, as well as an ethnographical one – which we applied from the very first inventory stage of the directive – with a quantitative approach monitoring the production of documents of objectives based on a national survey. The first difficulty of such a task involves following the local construction of documents of objectives while evaluating the creation of a network of expertise at a national level. We used the concept of 'local scenes' to convey this move towards action and its local basis without losing sight of its national and European dimension. Our proposal is, after the fashion of a group of geographers led by J. Levy, to 'refrain from contrasting territory and network, but to use the two terms as a conceptual couple'.[5] The construction of a level of European action, along with the production of locality, should be envisaged in this context as two processes feeding the new environmental policies. Each member state is, in fact, free to use the method of its choosing, provided it complies with an obligation to achieve results. The French interpretation of the directive falls within the framework of this general move towards the end of state monopoly in favour of multiple actors taking part at different levels, even though the rules governing local consultation procedures remain in the hands of the central authorities, i.e. the DIREN and the operators. The 'local scene' is this social space of collective creation aimed at action, which lies somewhere between instruction and deliberation, and includes stakeholders who are involved to varying degrees in local problems and networks. It is a key part of the process of relocating public action, where proximity practices and network practices come together. On the other hand, we can question how this new apportioning of the rural space will modify the relationship between a society and its territory.

The territorial dimension of Natura 2000 calls for diversifying fields of observation, not only in terms of habitats but also in terms of the regional socio-political context. The GRENAT researchers have worked in the four regions mentioned earlier, closely monitoring the creation of several DOCOBs. We strive in our research to define the diversity of forums for deliberation and to identify the debate structures set up, i.e. the places and forms of consultation and negotiation intended to result in stable agreements. Depending on the stage of the debate and the specific

---

[5] Estienne Rodary (2001) mentions Jacques Levy as the geographer who was behind these propositions. He defines territory as a bounded social space where the distances are continuous and exhaustive, whereas networks refer to fragmented social spaces with discontinuous distances.

layout of the different sites, the local history and overall social context influence the discourses and relationships established within the local scenes to a greater or lesser extent. We have approached the sites as areas of confrontation with a specific history, while being subject to the influence of a nature management policy of international dimensions. Furthermore, in order to define the nature of this forum for deliberation and the content of exchanges, we have worked within the framework of pragmatic sociology, so as to describe the interaction between stakeholders and to understand its effect.

This kind of qualitative approach cannot easily be applied to all of France due to both technical reasons and scientific interests. To overcome this difficulty, we complemented our research with a quantitative survey of the processes observed at the macro-sociological level. Our database was progressively developed as the documents of objectivess for each site were assembled. Its purpose was to make it possible to define the evolution of responsibilities in the creation of documents of objectives, to identify the groups in charge of their implementation, and to compare them with the pre-existing network of administrators. To achieve this, a postal survey was conducted involving the operators or project coordinators (Box 16.1) responsible for documents of objectives. Questions in the survey focused primarily on the resources and people mobilized by operators during the procedure. Potential conflicts, debates, negotiations, and disagreements are difficult to access when collecting information this way, and were therefore monitored locally at the site level.

Box 16.1 The GRENAT postal survey

Survey: The launch of a DOCOB is marked by the appointment of an operator. We send the operator a letter as soon as the creation of the DOCOB is fully underway. It contains a questionnaire to be returned to us, an explanatory letter and a letter of request from the Ministry of the Environment. If necessary, we also send a reminder two months later and make a follow-up call. Everything is done to maximize the likelihood of receiving a reply from the operator.

Questionnaire: It contains around 30 questions, many of which leave room for open answers. It is structured around nine sections: the operator; the site; the creation of the DOCOB; knowledge of the site and the definition of objectives; the steering committee (COPIL); the working groups; information and public consultation; difficulties; and the future.

Mailing letters and return rates: The first questionnaires were sent out in September 2001 and the last taken into account as part of this research on 18 February 2005. Of the 473 letters sent out, we received 326 replies corresponding to 363 observations, or a return rate of 77%.

The materials collected: Replies to the questionnaires but also materials concerning the operators, the composition of steering committees, geographical information, etc.

The database: Managed using ACESS and data processing using SAS 8.0. Each closed question has been the subject of descriptive statistical analysis and each section has been used to produce a synthesis. The open questions were analysed using PROSPERO. The geographical information on the sites was processed using MAP INFO 7.0.

Throughout the course of this chapter, we will examine certain elements of the morphology of local scenes and the interactions they generate, using data collected during these surveys (Pinton et al. 2007).

16.5 The Local Scenes of Biodiversity

16.5.1 The Steering Committee: An Official Consultation Body

The creation of a DOCOB is accompanied by a profusion of meetings in which different views of nature and nature management methods are on the agenda. The operator who is in charge of the DOCOB is responsible for organizing these different collectives. Within these diverse negotiation forums, the steering committee (known as the COPIL) represents the official consultation body. Its role is to supervise the creation of each document of objectives. The principal purpose of the meetings is to ensure the validation of all the measures and guidelines drawn up by the operators in relatively close cooperation with all the partners concerned. Its members are legal entities, or sometimes individuals, considered by the prefect and/or the DIREN as key partners in the management of the site. Members include, among others, representatives of state public services and institutions, regional authorities, professional organizations, environmental NGOs, organizations representing other users of the natural habitat, holders of land rights and users of resources, as well as the operator and those in an official capacity, such as the project coordinator. Based on documents provided by operators in response to our postal survey, we analysed the composition of 242 steering committees that oversaw a range of diverse habitats located throughout the entire country (Pinton et al. 2007: 111). The first finding was significant variations in the number of members. A total of 5,675 people were counted, giving an average of 28 members per COPIL (Box 16.2). In certain regions, the COPILs are fairly open, as in Ardèche, where there are many participants, who are likely to change over time, or the Marais Poitevin, where the COPIL sometimes involves up to 180 people. Other COPILs, however, are much smaller and established according to relatively strict standards, as in Nord-Pas-de-Calais or Burgundy.

An important reason for these variations is significant differences in the sizes of the sites, which affects the size of the steering committees. The representation of elected officials, such as mayors, councillors, and community chairs of communes (*communautés de communes*), depends on the size of the areas covered by the

Box 16.2 The steering committee: number of members

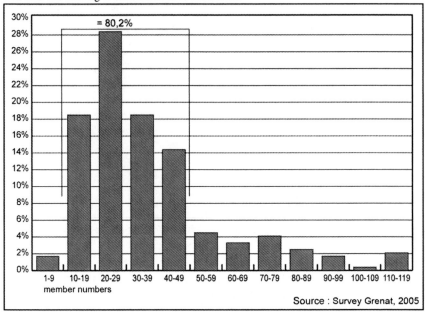

document of objectives. It is also clear that the natural environment of the habitats is influential: the more complex and diverse the natural habitat, the greater the number of interests represented. While a site composed of a homogeneous forested area will have a steering committee limited to a few government representatives, one or two elected officials, representatives of the forest (ONF, CRPF), hunting groups (ONCFS), and perhaps a representative of an environmental association, a composite site will also include additional authorities and representatives of farmers, amateur or professional fishermen, and industrial, tourism and sporting interests.

However, the differences noted above are insufficient to explain all of the variations observed in numbers. In general, the composition of the COPILs raises a number of questions, such as what are the designation criteria for its members and how representative are they? The answers to such questions differ from region to region, or even from place to place. If we limit ourselves to the national network, it is systematically structured around four colleges of unequal size: the government, local authorities, owners and users, and environmental protection representatives. However, their titles and outlines differ according to the orders of the prefect, there being no standard plan for steering committees to produce these orders.

We have therefore attempted to delve deeper into the analysis of the forms of representation within the COPILs (Box 16.3). The stakeholders are grouped according to broad types of activity. By pinpointing their presence, each group is counted as a single entity, regardless of the number of representatives attending. The following table shows that the most highly represented categories in the steering committees

Box 16.3 Representations of activities within the steering committees

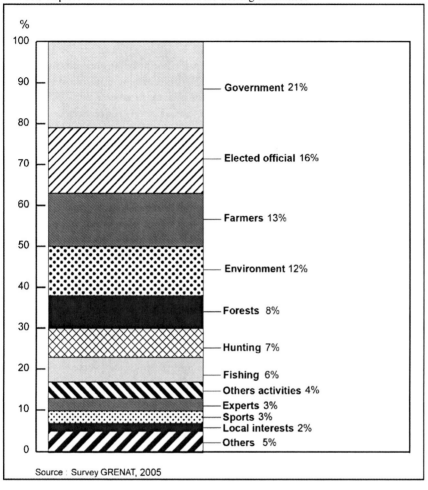

Source : Survey GRENAT, 2005

are the government, elected officials, farmers, the environment, forests, hunting, and fishing. Furthermore, the results confirm the low representation of other categories, especially those of scientists and experts.

A closer look at these data enabled us to demonstrate that environmental activities and interests are represented in a very heterogeneous way, as opposed to the other main types of activities represented by recognized spokespersons, such as the government with the DIREN (environment) and the DDAF (agriculture), the regional authorities with town halls and departmental councils (*conseils généraux*), farming with the chambers of agriculture (and also unions), forests with the CRPF and the ONF, hunting with federations and the ONCFS, and finally fishing, also with federations and the CSP (French Fishing Regulatory Body). The environmental institutions most represented in the COPILs are the regional (or departmental) conservatories of sites, but these are present in less than half of all steering committees. Next most represented are departmental protection associations and the Ligue pour la Protection des Oiseaux (LPO, League for the Protection of Birds). Compared with other major activities, environmental representation therefore appears to be relatively fragmented.

As already stated, the role of the operators and their project coordinators in these mechanisms is essential. Our results clearly show the highly diverse profiles of operators. Grouping operators into eight categories of activities makes it possible to single out three principal affiliations, regional authorities, environmental associations, and forests, which total over 80% of the DOCOBs. Operators thus play the role of mediator between science and action. They must create new representations of nature and redefine the space by integrating ecological and social criteria. They have three objectives: gathering and linking different knowledge registers (scientific, technical, and empirical); coordinating collectives, forums for debate, and meeting places of varying degrees of formality that take the form of information groups, working groups, and technical secretariat; and finally ensuring that communication is not neglected. These different sides to their work are intended to make it possible to turn forms of knowledge into consensual management measures that can be appropriated at the local level. Faced with the wide range of skills needed to meet these different demands, we have noted that relational activities play a strategic role that extends far beyond institutionalized collectives such as steering committees and working groups.

The data collected by GRENAT confirms, as we have just seen with the example of the steering committees, that the creation of DOCOB encourages stakeholders to work together – even those stakeholders who were not accustomed to doing so. In a complementary way, Natura 2000 contributes to supporting the move towards integrating new stakeholders into regional administration, including regional authorities that have not previously been involved in management, whether associations of communes, communities of communes, local authorities, certain environmental associations, or research consultancies. The local scenes are, in fact, forums in which new questions about environmental matters can be debated. The quality and true scope of the agreements reached within these collectives must now be appraised, and this is our next task.

16.5.2 Social Interaction Relating to Environmental Matters

Analysis of the phenomena of interaction within the 'local scenes' was carried out using material collected during field surveys conducted in the four regions. It is impossible to account for the full diversity of the debates and kinds of interaction that we witnessed. We therefore chose to analyse several situations in depth by working from a monographic approach. Situations of interaction usually take place within working groups where, in contrast to the practice in steering committees, different agents identify questions, evaluate their relevance, list their experiences, knowledge, and know-how, and attempt to provide a response in terms of action, which the specifications will take into account. The assumption made by GRENAT is that the creation of an document of objectives consists of reaching a compromise that is acceptable to public representatives and other stakeholders involved in the procedure, whether they are defending existing practices or promoting conservation objectives. This is a long-term and not always linear phase, in which a problem is first raised and then progresses through circles that usually extend beyond the working group, before returning to it in a form different to its original wording. Sometimes, exchanges are limited to forms of negotiation where stakeholders agree to modify their practices in return for compensation. Sometimes, the exchanges take on more complex forms and, in certain cases, lead to the development of a true dialogue that ensures the structures and constraints of all concerned are integrated, which results in forms of mutual learning.

In his analysis of the Marais Poitevin (Poitou-Charentes), Billaud focused on describing different situations of interaction and their impact on the mobilization and possible production of knowledge (Pinton et al. 2007: 130). The site observed is unusual in that it is very large and its management depends on a highly complex administration system in which several kinds of measures are superimposed on each other, especially where agriculture is concerned. The number of discussion forums increased and took the form of geographical working groups that are themselves broken down into thematic working groups, on the one hand, and a scientific committee comprising technical sub-groups, on the other. These different gatherings, involving up to 70 participants, also include cantonal meetings and even meetings at the town hall between project coordinators and farmers, in order to validate the cadastral mapping of land use. These apparently identical nature forums did not involve the same processes in terms of consultation and, more broadly speaking, of confrontation between participants. Moreover, they did not progress as a chain of necessarily cumulative interactions but offered different registers of interaction within working groups: agreement, debate, and uncertainty.

The knowledge-sharing system at the heart of the DOCOB construction process can also be analysed by monitoring a given problem from start to finish in the different local scenes, the treatment it receives, the interaction taking place in relation to it, and the way in which it is remodeled. We applied this method to a clearly defined issue in Burgundy, that of the management of the Puisaye Ponds (Pinton et al. 2007: 143). The steering committee is a place for formulating diverse questions and queries concerning the procedure, its consequences, and the methods used for its implementation, while also revealing the resulting fears, uncertainties, and differences in opinion. We show that the role of technical meetings concerning ponds is more concerned with defusing conflicts than solving all stated problems. The meetings appear as an opportunity for explanation, including clarification of technical and scientific terms, and comparison of proposals based on individual experience. The exchanges that take place between operators and administrators, owners, or users of the ponds make it possible to pinpoint problematic measures and to establish contacts that will prolong the meeting. Individual meetings, telephone exchanges, and emails are used to polish certain measures within proposals and to arrive at a lawful expression that takes local issues into account.

The abovementioned examples show that such interaction situations are capable of shedding new light on social and natural issues and, consequently, of redefining the obligations that exist between concerned persons and natural entities. Debate on the issue of practical knowledge and know-how, the resulting agreements and disagreements, and emerging uncertainties all constitute a specific political process aimed at establishing new forms of organization, admittedly under the aegis of the public authorities, but also affected by the interactions organized by the procedure. If there is, indeed, a learning process, it is at the heart of these new cognitive and practical approaches.

16.6 What Kind of Legitimacy Do Environmental Policies Have?

Natura 2000 is an example of the progressive increase in society's reaction to environmental issues. Our observations highlight two forms of the relationship between knowledge and action identified within the procedure. The first phase of the directive's application (drawing up inventories and creating national lists) was based on a belief in a radical separation between science and action. Scientific knowledge was seen as stable and considered independent of action and was then to be translated into public action. The territorial testing of these two opposing worlds led to a re-think of political action towards an approach based on deliberation and participation, which gradually proved to be a source of respite. The issue of the public legitimacy of this once highly controversial directive gave way to an important process of re-defining the biodiversity to be protected. The procedural approach that was applied to the stage of drawing up documents of objectives has proven even more suitable when the main challenge is to take into account complex problems and scientific uncertainties. The link that is established between knowledge and action depends primarily on the competence of the operator, who is responsible for organizing debates, and on the efficient organization of the steering committee.

In this sense, our local scenes correspond to what Habermas defines as the place for 'legitimate communicational power', in that they enable stakeholders from all different sectors to reach agreements, even if the principle is imposed externally. The agreements actually correspond to compromises between European demands, scientific data, and local practices. We can assume that these local scenes foreshadow public actions that should accompany the environmentalization process in rural France, as well as at a European level.

The monitoring of procedures by GRENAT, whether at the local or national level, makes it possible to outline the principal characteristics of this environmental policy in the making: a strong institutional dimension (European framework), but also a good deal of room for flexibility (local deliberation procedures and negotiations, etc.), knowledge sharing (field testing and the re-creating of knowledge), and finally a joint construction process capable of blurring the initial positions (redefinition of issues or practices, reconstitution of different groups of stakeholders). It is especially remarkable that we are seeing new stakeholders such as fishermen and landowners taking the stage.

While not totally new, the scope of this kind of policy is unusual in terms of its scale and the complexity of issues raised. We can, nevertheless, question the real impact of the agreements reached. In other words, does this approach lead to lasting changes in the management of nature, and does it generate new configurations that are seen as essential to a biodiversity conservation policy?

Despite a number of innovations, our research leads us to conclude that the forces at work in the environmentalization process of the French territory are relatively weak; the changes implemented remain especially fragile and understated. We have already highlighted the inadequacy of environmental representation within the different collectives along with the very low involvement of scientists at this stage of the procedure. For the moment, there is no apparent structure that is capable of collecting the contributions of this social experiment within the environmental sphere, which is itself highly segmented. Furthermore, the administrative complexity of assembling the DOCOBs and the meagre financial incentive provided to the contracting parties constitute a major obstacle to prolonging and consolidating the work undertaken thus far.[6]

One of the reasons for this fragility is that biodiversity conservation, unlike other issues such as conserving water resources, falls within the category of environmental problems that are part of an unstable or controversial sphere. On the one hand, controversy develops because the assessment of biodiversity and its dynamic remains difficult from a biological point of view. Controversy also develops because different registers of legitimacy are applied depending on the objectives of the proposed protection. The use of the concept of biodiversity is then difficult to grasp, especially for stakeholders in the field, and it clearly does not encourage their involvement.

Nevertheless, while at the time of its application in 1992 the directive had no legitimacy in France, biodiversity conservation has been gradually imposed from the outside. Recent developments in international negotiations concerning the Convention on Biological Diversity (CBD), represented in France by the *Stratégie Nationale pour la Biodiversité* (National Strategy for Biodiversity), reflect the beginnings of an important change in the system of reference for public action, change that has to be continued.

References

Alphandéry, P. and F. Pinton (1999). 'L'application de la directive en région Bourgogne.' In Rémy et al., pp. 232–256.

Alphandéry, P. and A. Fortier (2001). 'Can a territorial policy be based on science alone? The system for creating the Natura 2000 network in France.' *Sociologia Ruralis* 41(3): 311–328.

Dodier, N. (2003). *Les leçons politiques de l'épidémie de Sida*. Paris, EHESS.

Godard, O. (2002). 'Du risque à l'univers controversé', in Séminaire du GIS *Du risque à l'univers*

---

[6] All the measures are not eligible. Those who commit themselves through contracts are accepting some constraints, which may or may not be linked to funding or compensation.

226   F. Pinton

controversé et au principe de précaution ou le raisonnable à l'épreuve. Risques collectifs et situations de crise, actes de la 19ème séance, Paris, Ecole des Mines de Paris.

Hervieu, B. and J. Viard (1996). *Au bonheur des campagnes*, Paris, éd. de l'Aube.

Pinton, F. (coord.), P. Alphandery, J-P. Billaud, C. Deverre, A. Fortier and G. Géniaux (2007). *La construction du réseau Natura 2000 en France. Une politique européenne de conservation de la biodiversité à l'épreuve du terrain*, La Documentation française, Paris.

Rémy, E. (coord.), P. Alphandery, J-P. Billaud, N. Bockel, C. Deverre, A. Fortier, B. Kalaora, N. Perrot and F. Pinton (1999). *La mise en directive de la nature. De la directive Habitats aux prémices du réseau Natura 2000*, Rapport pour le MATE. Paris.

Rodary E. (2001). 'Perspectives territoriales et contrôle réticulaire dans les politiques participatives de conservation de la nature en Zambie et Zimbawe.' *Geocarrefour* 76(3): 225–232.

Valentin-Smith, G. et al. (1998). *Guide méthodologique des documents d'objectifs Natura 2000*. Quétigny, Réserves Naturelles de France/Atelier Technique des Espaces Naturels.

List of Abbreviations

COPIL	Comité de Pilotage (Steering Committee)
CRPF	Centre Régional de la Propriété Foncière
CSP	Conseil Supérieur de la Pêche (French Fishing Regulatory Body)
DDAF	Direction Départementale de l'Agriculture et de la Forêt (Departmental Directorate of Agriculture and Forests)
DIREN	Direction Régionale de l'Environnement (Regional Directorate for the Environment)
DOCOB	Document d'Objectifs (document of objectives)
GRENAT	Groupe de Recherches sociologiques sur la Nature (Sociological Nature Research Group)
IFB	Institut Français de la Biodiversité (French Institute of Biodiversity)
LPO	Ligue pour la Protection des Oiseaux (League for Birds Protection).
MATE	Ministère de l'Aménagement du Territoire et de l'Environnement (French Ministry of National Development and Environment)
MEDD	Ministère de l'Ecologie et du Développement Durable (French Ministry of the Environment and Sustainable Development)
ONCFS	Office National de la Chasse et de la Faune Sauvage (National Office of Hunting and Wildlife)
ONF	Office National de la Forêt (National Forest Office)
PACA	Région Provence-Alpes-Côte d'Azur
ZNIEFF	Zones Naturelles d'Intérêt Ecologique, Faunistique et Floristique (Natural Areas of Ecological and Wildlife Interest)

Chapter 17
The Local Implementation of Nature Policy

Deliberative Democracy or Interest Politics?

Henk van den Belt

Florence Pinton and some of her colleagues from GRENAT like Pierre Alphandéry and Agnès Fortier have written extensively about the implementation of the EU Habitat Directive and the creation of the Natura 2000 Network in France. I would like to venture a concise and highly selective summary of their work here. From 1992 until now the French attempt at implementation can be divided into two radically different periods. At first, the selection of the sites for protection and the delineation of their boundaries, which was thought to be a purely scientific exercise, were set up following a top–down approach. Consultation of the regional and local rural actors was planned as a second stage in which only the appropriate management measures for the selected sites would be discussed (cf. Pinton 2001). However, fierce opposition from regional authorities and rural groups – foresters, huntsmen, farmers, fishermen – already during the first stage of the implementation process forced the French Ministry for the Environment to abandon this top–down approach. In 1996 the implementation was suspended. When the implementation was re-launched in 1997, the new policy initiative placed much more emphasis on consultation and consensus-seeking. However, this more deliberative approach seems to have resulted in a 'dilution' of the nature goals to be realised. One outcome has been a drastic reduction of the surface area of the sites to be notified to the European Union (from 7 million ha or 13% of the national territory in March 1996 to 3,144,000 ha or 5.7% of the national territory in 2000). Existing human activities using nature have also found recognition. Some French ecologists complain about the 'denaturation' of the inventory work (Alphandéry and Fortier 2001: 320). Due to the limitations of her sociological approach, Dr. Pinton declares herself incompetent to judge 'the soundness of choices with regard to the environment' and confines herself to dealing with the question of social equity. I find this self-imposed limitation rather unfortunate, as the substantive outcomes of nature policy are precisely the object of normative evaluation. At any rate, it is hard to dispute the overwhelming impression that

Henk van den Belt
Department of Applied Philosophy, Wageningen University & Research Centre
E-mail: henk.vandenbelt@wur.nl

J. Keulartz and G. Leistra (eds.), *Legitimacy in European Nature Conservation Policy:*
Case Studies in Multilevel Governance, 227–230. © Springer 2008

over time the nature goals to be realised have been considerably watered down compared to the ecologically more ambitious targets that had been chosen at the outset.

By and large, this story is not much different from what happened in The Netherlands and perhaps also what happened in other European countries (see, for example, van den Belt 2004). In the Netherlands, the realisation of a National Ecological Network has also often met fierce local resistance. To appease farmers and other inhabitants of rural areas, public consultation has been deployed to arrive at revised, socially shared goals for nature protection. The effect of such exercises has been a general 'dilution' of the nature goals to be realised, i.e. a shifting towards lighter shades of green (e.g. towards 'semi-natural nature' and 'multi-functional nature', according to the division of main types of nature in the national classification system). At one time (in 2000), André van der Zande, the inventor of the idea of a National Ecological Network, remarked that 'democracy' might be a precious good, but that nature has had to pay a very high price for it. However, it remains to be seen to what extent consultation processes with local stakeholders deserve to be called 'democratic'.

We are confronted here with what seems to be a very serious dilemma. On the one hand, we have to recognize that a top–down policy does not work. On the other hand, the attempt to involve local stakeholders like farmers, foresters and hunters in setting the goals for nature policy through public participation tends to result in outcomes that are much less satisfactory from an ecological point of view.

How to deal with this dilemma? I want to raise a few questions to clarify the underlying problem.

One might hold that science, in this case ecological science, has no monopoly on defining which sites, habitats and species are worth protecting. Dr. Pinton suggests that local stakeholders questioned the credibility of the science used in making the preliminary inventories of the natural heritage in France and that they advanced the claim that their own 'local knowledge' (or practical knowledge) was more adequate. But how credible are such claims made by local stakeholders, given that they are obviously raised for opportunistic reasons in the context of a fierce power struggle? Today it is indeed fashionable in many branches of social science to oppose the notion of 'local knowledge' (or 'indigenous knowledge') as a valuable asset against abstract, formal and presumably impractical scientific knowledge, but perhaps this romanticist flirtation with local knowledge is highly irresponsible and amounts to no more than simply taking sides with rural interests against the overall public interest of nature protection.

I suspect that the actual process of deliberation and participation is depicted in far too rosy colours when Dr. Pinton describes it as a 'learning process' and as 'an important process of redefining the biodiversity to be protected' (p. 224). She even states:

"... . [O]ur local scenes correspond to what Habermas defines as the place for 'legitimate communicational power', in that they enable stakeholders from all different sectors to reach agreements, even if the principle is imposed externally. The agreements actually correspond to compromises between European demands, scientific data, and local practices." (p. 224)

I think that it may be highly misleading to invoke notions of deliberative democracy in referring to these local scenes. As the Australian political scientist John Parkinson also remarks, there is often 'very little in mass public communication, including a great deal of media debate, large-scale referendum processes, or even public meetings, which merits the label "deliberative"' (Parkinson 2003: 181). In fact, the criteria for genuine deliberation are extremely demanding: participants must exhibit communicative competence, reciprocity, and the willingness to be persuaded and to set aside strategic concerns and behaviour as well as meet the condition of inclusiveness (Ibid.). Pointing to the redefinition of biodiversity as the outcome of a learning process among the participants will not do to prove the transformative character of the allegedly 'deliberative' process that is often highlighted by the advocates of deliberative democracy, as in this case it is not so much the preferences of the participants themselves as the official definition of biodiversity that has been transformed. In my opinion, it would be much more appropriate to describe the processes occurring in the French consultations in terms of bargaining and negotiation. Using the term 'deliberation' falsely suggests that what counts is primarily the force of arguments rather than the weight of interests. As Ian Shapiro reminds us in his criticism of some ideals of deliberative democracy, most politics still centres on interests and power (Shapiro 1999). With regard to the criterion of inclusiveness one can also raise serious doubts about the democratic character of the whole process. After all, city-dwellers may also have a strong interest in the conservation of natural biodiversity, but to what extent are they included in regional and local consultations? If local stakeholders are allowed to 'redefine' the biodiversity to be protected in the course of the consultation process, are there any guarantees left that the final outcome will be acceptable to the urban population? Looking at numerical shares: How democratic is a process in which some 90% of the population that happens to live in cities and towns is actually disenfranchised from taking part in the decision-making process? How democratic is a process in which the fate of nature and biodiversity is placed in the hands of those 10% of the population who happens to live in the countryside?

In France as well as in the Netherlands, farmers, hunters, foresters and fishermen often see themselves as the authentic 'owners and users of nature' (Alphandéry and Fortier 2001: 316) and resent any intrusions from Europe or the national government in their God-given rights. If property rights is indeed the name of the game, this would suggest that the appropriate instruments of nature policy are power and money rather than arguments and deliberation. Already in 1990, the economists David Bromley and Ian Hodge argued that the prevalent regime of property rights in European agriculture is such that farmers will be persuaded into more socially desirable modes of production (e.g. modes that are more respectful of nature, landscape and environment) only if they are offered enough 'bribes' by the state (Bromley and

Hodge 1990). On this view, a drastic reform of the regime of property rights in the countryside may indeed be long overdue.

References

Alphandéry, P., and A. Fortier (2001). 'Can a Territorial Policy be Based on Science Alone? The System for Creating the Natura 2000 Network in France'. *Sociologia Ruralis* 41: 311–328.

Belt, H. van den (2004). 'Networking Nature, or Serengeti Behind the Dikes'. *History and Technology* 20(3): 311–333.

Bromley, D., and J. Hodge (1990). 'Private property rights and presumptive policy entitlements: Reconsidering the premises of rural policy'. *European Review of Agricultural Economics* 17: 197–214.

Parkinson, J. (2003). 'Legitimacy Problems in Deliberative Democracy'. *Political Studies* 51: 180–196.

Pinton, F. (2001). 'Conservation of Biodiversity as a European Directive: The Challenge for France'. *Sociologia Ruralis* 41: 329–342.

Shapiro, I. (1999). 'Enough of Deliberation: Politics is about Interests and Power'. In Stephen Macedo (ed.), *Deliberative Politics: Essays on Democracy and Disagreement*. New York/Oxford: Oxford University Press, pp. 28–38.

The Oder Valley in Brandenburg (by Markus Leibenath)

Chapter 18
Legitimacy of Biodiversity Policies in a Multi-level Setting

The Case of Germany

Markus Leibenath

18.1 An Introduction to Perception and Acceptance of Nature Conservation in Germany

In 2004, a large German weekly newspaper published an extensive article on Natura 2000 under the heading 'The Hamsters' Revenge' (Frömmel 2004: 63). The article dealt with a population of hamsters (*Cricetus cricetus*) that prolonged the planning process for an industrial zone on the outskirts of the city of Mainz for more than a year and led to additional costs of some 70 million Euros for relocating the animals. The hamsters had to be relocated because they are listed in annex IV of the Habitats Directive as an 'animal species of Community interest in need of strict protection'. The German media (e.g. Berger 2002: 6) have reported on several conflicts between Natura 2000 and land-use interests, on the lack of official information with regard to Natura 2000 and on insufficient participation of local stakeholders in the processes surrounding Natura 2000.

However, the image of nature conservation as an obstacle to economic and social developments is not restricted to Natura 2000 but also applies to other instruments of nature conservation in Germany. In most of the country's large protected areas like nature parks, biosphere reserves, and national parks, there have been and still are fierce confrontations about the pros and cons of conservation and the different approaches to it (Stoll-Kleemann 2001: 32–38).

Most German researchers who analyse social conflicts caused by the implementation of biodiversity policies have focused on the dimension of 'acceptance'. They have studied the degree to which different groups accept certain conservation measures and the factors that influence acceptance. Furthermore, they have proposed

Markus Leibenath
Department for Regional Development and Landscape Ecology,
Leibniz Institute of Ecological and Regional Development (IOER), Dresden, Germany;
Dresden Technical University
E-mail: M.Leibenath@ioer.de

J. Keulartz and G. Leistra (eds.), *Legitimacy in European Nature Conservation Policy:*
Case Studies in Multilevel Governance, 233–250. © Springer 2008

strategies and measures to raise the acceptance of biodiversity policies among local and regional stakeholders (e. g. Heiland 1999; Leibenath 2001: 19–25; Rentsch 1988; Sauer et al. 2004, SRU 2002: 25–58; Stoll 1999; also cf. Toogood et al. 2004).

Acceptance is defined as an individual's positive, affirmative attitude towards an object. It can develop as an unintended by-product or as an explicit goal of certain actions. Acceptance can be an end in itself or it can represent a precondition for further interaction (Job 1996: 160; Lucke 1995: 80, 88 and 103; Rentsch 1988: 10, 12).

The notion of acceptance is closely linked to the term 'legitimacy'. In a sense, acceptance is the counterpart of legitimacy. It designates the actor-centred, subjective reverse of legitimacy. It highlights the subjective component of social and legal norms and institutions that could not obtain comprehensive legitimacy in temporary democratic societies without acceptance (Lucke 1995: 93). Boedeltje and Cornips (2005: 15) make a similar distinction when they refer to Easton's attitudinal categories 'diffuse support' and 'specific support' for measuring the legitimacy of political institutions.

According to Scharpf (2004: 3), legitimacy is a functional prerequisite of forms of governance that are simultaneously efficient and liberal. Thus, legitimacy, unlike acceptance, is related to the character of policy or government. Legitimizing arguments 'establish a moral duty to obey [...] collectively binding decisions even if they conflict with individual preferences' (Scharpf 1998: 2).

Three dimensions of legitimacy can be identified. The first one is input legitimacy, which implies that collectively binding decisions should originate from the authentic expression of the preferences of the constituency in question. Such authentic participation implies the possibility of articulating consent to or dissent against proposed policies and of influencing discussion on these proposals (Getimis & Heinelt 2004: 10; Scharpf 1998: 2).

The second dimension is output legitimacy, which centres on the effectiveness and efficiency of governance activities with regard to the achievement of issue-specific goals, but also focuses on the capability of fulfilling more general functions that go beyond issue-specific problem-solving. Both the democratic system as a whole and any democratic decision are measured according to the degree to which they solve prevalent problems. The output dimension of democratic legitimacy is linked to the assumption that collectively binding decisions should serve the common interest of the constituency (Getimis & Heinelt 2004: 10; Scharpf 1998: 2; Wolf 2002: 17).

The third dimension is called throughput legitimacy and refers to the quality of governance procedures in which actors are involved. The main criterion here is transparency because citizens have to understand how measures are taken and who is responsible for them in order to make actors accountable for what they have done and to make them understand what the alternatives that have to be decided upon were. Hence, accountability forms another criterion in this dimension of legitimacy. Throughput legitimacy of political systems or of individual decisions will be high if activities are transparent and take place in public instead of behind closed doors, if they show a high deliberative quality which is based on arguing instead of bargaining, if they are responsive and reliable, and finally if the decision-makers are identifiable, responsible and accountable (Getimis & Heinelt 2004: 10; Wolf 2002: 17, 25 and 27).

The need to legitimize a policy or a decision depends on the underlying actor constellation, which can be analysed from the perspective of game theory. Scharpf (2004: 4) presents a typology of theoretical actor constellations, ranging from pure coordination games via coordination games with distributive conflicts and dilemma games to zero-sum games. The latter cause the strongest need for legitimizing arguments because here the interests of one group can only be met at the expense of those of another. The group of 'losers' will not be reconciled by pointing to their general preferences or to their basic self-interest because they are clearly the losers in every respect. The constellation of a zero-sum game can be found in many conservation conflicts. Usually, the conflicts can be analysed as distributive conflicts because relatively few local stakeholders are asked to comply with strict rules, to accept restrictions for land-use and so on for the sake of public welfare without some kind compensation (Möller 1995: 29).

One should realize, however, that, insofar there is a 'crisis of legitimacy' of biodiversity policies in Germany, this is only true with regard to persons or groups who are relatively closely affected by conservation projects. Opinion polls have revealed, for instance, that a seventy percent majority of the German population is in favour of additional national parks in Germany and that ninety-five percent think national parks are important (Umweltstiftung WWF Deutschland 1998). These polls present a picture that differs from the reports of fierce protests in local communities against conservation (Leibenath 2001: 21–23).

At this point, it should already be clear that legitimacy of top–down policy-making in the field of biodiversity policies in Germany is ever decreasing. The intention of this chapter is to analyse whether this has led to an increased emphasis on procedural legitimacy in the processes of German nature conservation policy formulation and implementation and what role the EU has played in generating this shift. The chapter also sheds light on the importance of national traditions and institutions, on trade-offs between different types of legitimacy, and finally on coordination problems between different agents and different policy levels. This is done on the basis of two empirical lenses or short case studies: one on the implementation of Natura 2000 in the Federal Land of Brandenburg and one on the establishment and management of the Müritz National Park.

Biodiversity politics in Germany has turned into a complex multi-level policy arena in the past few years, or more accurately, into a complex set of interlinked horizontal and vertical policy arenas at different levels. Traditionally, nature conservation is decentralized and falls into the competence of the Federal Länder. However, the Länder and the federal government also share some competences. With the introduction of a European biodiversity policy, a third level has been added. This new level has created a completely novel situation for the conservation authorities at the Länder level: while they used to be autonomous to a very large extent, the conservation authorities now have to coordinate the selection and delineation of protected areas with the Federal government and with European institutions. As Benz (2000: 34) points out, this change has resulted in a new type of dependency because European decisions and regional implementation are now linked in a hierarchical pattern: higher-level decisions set the framework for lower-level decisions and regional actors are not able to veto European decisions.

Against this backdrop the next section gives an overview of German biodiversity policies and the structure of the respective policy arenas as well as of the legal and practical implementation of Natura 2000 in Germany. Next, the third section includes the two case studies. In the fourth and final section, the main findings are discussed and some conclusions are drawn on the initial research questions.

18.2 Overview of German Biodiversity Policies

18.2.1 Institutions and Actors in the Biodiversity Policy Arena

In theory, nature conservation falls into the competence of the Länder. This means that there are, in practice, significant differences in the conservation policies of the sixteen Länder. However, according to article 75, paragraph 1 of the Basic Law for the Federal Republic of Germany (similar to other countries' constitutions), the federal government has the power to enact provisions on nature conservation and landscape management as a framework for the Länder legislation, provided that 'the establishment of equal living conditions throughout the federal territory or the maintenance of legal or economic unity renders federal regulation necessary in the national interest' (article 72, paragraph 2 of the Basic Law of the Federal Republic of Germany).

The federal government has made use of its power and passed the first Federal Nature Conservation Act in 1976 (last revised in 2002). In addition, every Land has its Nature Conservation Act, which fills the framework established by federal law, e.g. the 'Hessisches Gesetz über Naturschutz und Landschaftspflege' (Nature Conservation and Landscape Planning Act for the German State of Hesse).

Framework legislation such as the Federal Conservation Act may only under exceptional circumstances contain detailed or directly applicable provisions (article 75, paragraph 2 of the Basic Law of the Federal Republic of Germany). In the case of the Federal Conservation Act, these exemptions concern, for example, European and international aspects of conservation like Natura 2000 or trafficking in endangered species and the recognition of conservation associations as well as the legal remedies available to them (article 11 of the Federal Conservation Act).

Protected areas are one of the major instruments of nature conservation in Germany. There is a broad spectrum of protected area types, including nature conservation areas, national parks, biosphere reserves, landscape protection areas, nature parks, natural monuments, protected components of landscapes and different forms of legally protected biotopes (articles 22–31 of the Federal Nature Conservation Act). Designation of protected areas falls under the exclusive competence of the Länder, except for the designation of national parks, where the Länder have to consult with the federal government (article 22, paragraph 4 of the Federal Nature Conservation Act).

Who are the main actors in the biodiversity policy arena in Germany? First and foremost, these include all the ministries for nature conservation at the federal and Länder levels. Most of these ministries have technical agencies at their disposal, e.g. the Federal Agency for Nature Conservation or the Bavarian Agency for Environmental Protection. The Länder governments have made different institutional arrangements for the devolution of competencies to lower levels of government such as county administrations ('Landkreise'). Conservation associations like the German Association for Nature Conservation (Naturschutzbund Deutschland, the German branch of BirdLife International) or the German Federation for Environmental Protection and Nature Conservation (Bund für Umwelt und Naturschutz Deutschland, the German branch of Friends of the Earth) are another important group of actors because they enjoy far-reaching rights in the process of issuing ordinances on nature conservation and in infrastructure planning procedures (article 58 of the Federal Nature Conservation Act).

With regard to the input legitimacy of biodiversity policies, it is important to know what rights of participation the broader public has. The Federal Nature Conservation Act stipulates only that:

'in the case of measures planned for nature conservation and landscape management, the early exchange of information with the parties concerned and interested segments of the general public shall be safeguarded' (article 2, paragraph 1, clause 15 of the Federal Nature Conservation Act).

The nature conservation acts of the German Länder are more specific in this regard. For instance, the Conservation and Landscape Planning Act for the German State of Hesse states in article 16 that if the conservation agencies in charge want to designate any protected area, they have to inform the landowners as well as affected public agencies ('Träger öffentlicher Belange'), including relevant NGOs, in advance. In doing so, both landowners and public agencies shall have the opportunity to express their opinions on the proposed plans. Hence, only a very formal participation procedure is legally required.

18.2.2 Implementation of Natura 2000 in Germany: A Tale of Neglect, Delays and Lawsuits

Generally speaking, one can distinguish between formal and practical implementation. 'Formal implementation' refers to legal aspects, e.g. the implementation of European directives into the national laws of the member states. By contrast, 'practical implementation' refers to the practice of government or governance, i.e. to the actions and interactions of implementing agencies and those who are addressed by a policy. Practical implementation may also imply altering administrative structures and procedures (Knill 2003: 164).

With regard to the formal implementation of the European Birds and Habitats Directives, the German federal government implemented both by amending the Federal Nature Conservation Act in 1998. According to article 23, paragraph 1 of the

Habitats Directive, Germany was required to 'bring into force the laws, regulations and administrative provisions necessary to comply with this Directive within two years of its notification', which would have been by 1994. Hence, there was a delay of four years.

The practical implementation of the two directives, especially of the Habitats Directive, consisted of three stages. The first stage, submission of a national list of proposed Sites of Community Importance (pSCI), had to be finalised by 1995. In the second stage, the European Commission had to establish a list of Sites of Community Importance (SCI) in consultation with the Member States by 1998. Once a site of Community importance had been adopted, the member state concerned had to designate that site as a special area of conservation (SAC) within six years, which would have been no later than 2004. This was the third and final stage (article 4, paragraphs 1, 3 and 4 of the Habitats Directive).

Actually, the three stages were not handled consecutively but partly simultaneously because Germany encountered many delays in the first stage and in the formal implementation of the European directives. Germany neither kept the deadlines set by the European Commission nor proposed a sufficient number of SCIs, although the country bears a special responsibility for the conservation of habitat types in the EU's continental biogeographical region. Germany only submitted the first pSCIs to the European Commission in 1996, which was followed by several 'tranches' until the end of 2001.[1] In 1998/9, the European Commission, in return, sued the German government at the European Court of Justice (ECJ), which convicted Germany in September 2001. In April 2003, the European Commission decided to launch a penalty procedure according to article 228 of the Treaty Establishing the European Community, but it did not make use of the opportunity to bring the case again before the ECJ in order to have a lump sum or a penalty payment imposed upon Germany (BfN 2005; Kehrein 2002: 5; SRU 2004: 113–120).

In 2005, Germany completed the site selection and designation process. Currently, approximately 13% of Germany's terrestrial surface area is protected as Natura 2000 sites. In particular, there are 3,500 SCIs, which cover 7.0% of the terrestrial surface area (EU-25 average: 11.6%). Moreover, there are almost 500 Special Protection Areas (SPA), based on the Birds Directive, which cover 6.4% of the terrestrial surface area (EU-25 average: 8.2%) and which partly overlap with the SCIs (BfN 2005; European Commission, GD Environment 2005).

The competencies and responsibilities for selecting, proposing and designating SCIs and SPAs are split between the German federal government and the Länder governments. The Länder have to select the sites to be proposed to the European Commission and to designate them, whereas the Federal Ministry for the Environment, Nature Conservation and Nuclear Safety has to notify and to consult on the selected sites with the European Commission (article 33 of the Federal Nature Conservation Act).

---

[1] The term 'tranche' refers to the different rounds of notifications of pSCIs to the Commission and the subsequent amendments.

The major reason for the delays and shortcomings in implementing Natura 2000 in Germany can be found in real or assumed conflicts with landowners and land-users, who were lobbying vigorously against Natura 2000. As the Länder had to bear the financial and legal consequences of selecting an area for Natura 2000, they followed a conflict-averse strategy in the first place and based their first tranche of pSCIs on political instead of ecological considerations. The Länder governments tried to select only areas that were already under protection. However, strictly protected areas like nature conservation areas and national parks covered less than three percent of Germany's terrestrial surface area and thus were definitely insufficient. Moreover, many existing protected areas were too small to become SCIs (SRU 2004: 125; Ssymank 2005: 26).

According to German case law, no public participation is necessary in the first stage of the Natura 2000 designation process in which the national governments prepare their lists of pSCIs and submit them to the European Commission. The argument for this is that to include a site in the national list of pSCIs does not make any difference as compared to the mere fact that a species or a habitat type of Community interest exists on that site. Some German landowners do not share this view and have filed an action at the ECJ's Court of First Instance. They claim that the Court should annul the Commission's decision to adopt the SCIs list for the Alpine biogeographical region. They submit that the contested decision infringes on their fundamental rights because no form of right of participation was bestowed upon the landowners concerned (OJ 24.07.2005). The Court's decision is still pending.

Notwithstanding any legal requirements, the Länder governments have practiced various forms of public participation in the process of selecting Natura 2000 sites. In addition to interdepartmental coordination, which was applied in all cases, there were three basic types of public involvement:

- Participation by public agencies and relevant NGOs ('Träger öffentlicher Belange'),
- informal participation by local stakeholders, chiefly in public hearings, on-site meetings and disclosure of lists of pSCIs, and
- formal participation by local stakeholders, including the disclosure of pSCIs lists as well as calls for submitting written objections.

The objective of the Länder was not so much to reach consensus because in most cases there was hardly any room for negotiation. Instead, the Länder wanted to obtain additional information about the pSCIs and to improve the quality of their pSCIs lists. However, the lists were only modified in a few cases. The following table gives an overview of the participation procedures that have been applied in the German Länder.[2]

The table underpins two findings. First, implementation procedures in German biodiversity policies differ significantly from one Land to another. Second, most Länder only applied more extensive approaches to public participation in the late phases of the Natura 2000 site selection process.

---

[2] This information is based on a telephone survey by Dipl.-Ing Alexandra Sauer (Freising), who generously agreed to have her findings used in this article.

Land	Type of public participation with regard to the different tranches of pSCI notifications		
	Participation of public agencies and relevant NGOs	Informal participation of local stakeholders	Formal participation of local stakeholders
Baden-Wuerttemberg			2nd and 3rd tranches
Bavaria			2nd and 3rd tranches
Brandenburg	2nd and 3rd tranches		
Hesse		4th tranche	
Lower Saxony	1st tranche	2nd tranche	
Mecklenburg-Western Pomerania	All tranches		
North Rhine-Westphalia	All tranches	All tranches	
Rhineland-Palatinate			
Saarland	2nd tranche	1st tranche	
Saxony		2nd tranche	3rd tranche
Saxony-Anhalt		All tranches	
Schleswig-Holstein			2nd and 3rd tranches
Thuringia		All tranches	

Fig. 18.1 Public participation in the process of selecting pSCIs in the German Länder (except for the 'city states' Berlin, Bremen and Hamburg)

18.3 Empirical Lenses on Strategies for Obtaining Legitimacy for Biodiversity Policies

18.3.1 Selection of Cases

The purpose of the individual cases or the empirical lenses is to show what strategies conservation agencies and policy-makers have applied to improve the input and throughput legitimacy of biodiversity policies. Because the implementation of nature conservation differs a great deal from one German Land to another, cases from different Länder have been chosen to illustrate the variety of implementation strategies employed in Germany. Because this book focuses on European Nature Conservation Policy, one case about the implementation of Natura 2000 has been selected. However, because the approach of Natura 2000 is unique, another case on national parks has also been included to illustrate a more traditional instrument of German conservation policy. The first case study deals with the implementation of Natura 2000 in the German Land of Brandenburg; the second with the establishment and management of the Müritz National Park in the Land of Mecklenburg-Western Pomerania.

18.3.2 Natura 2000 in the Federal Land Brandenburg[3]

Brandenburg is the Land that surrounds the German capital Berlin, which, however, does not belong to it. The capital of Brandenburg is Potsdam, southeast of Berlin. The Land has an area of 29,477.16 square kilometres and about 2.6 million inhabitants. The countryside is characterized by a large number of rivers, canals, and lakes. All in all, Brandenburg has 33,000 kilometres of rivers and approximately 3,000 lakes, which makes it the Land with the greatest number of water bodies (Land Brandenburg 2005).

Brandenburg hosts 36 habitat types from annex I of the Habitats Directive, 9 of which are 'priority habitat types', and 44 species from annex II, 2 of which are 'priority species' (Zimmermann et al. 2000: 45).

The Habitats Directive was formally implemented in Brandenburg when the 'Conservation and Landscape Planning Act for the Land Brandenburg' was adopted in 2004, six years after the Federal Nature Conservation Act had been passed. The Ministry for Rural Development, Environment and Consumer Protection is officially responsible for selecting and notifying Natura 2000 sites, whereas the Environment Agency of Brandenburg is in charge of the day-to-day running of the site and, in fact, does all the technical work.

According to article 26b, paragraph 1 of the Conservation and Landscape Planning Act for the Land Brandenburg, SPAs and SCIs have to be designated as national parks, nature conservation areas, landscape protection areas, or protected components of landscapes. Other forms of protection are also possible, for example, on the grounds of voluntary contractual agreements with landowners. The municipalities and public agencies that are affected by the designation of such protected areas have the opportunity to comment on proposed protection ordinances. In order to do this, draft versions of protection ordinances are accessible to the public for at least one month. All objections that have been submitted by citizens affected or by public agencies have to be answered by the nature conservation agency in charge (article 26b and 28 of the Conservation and Landscape Planning Act for the Land Brandenburg).

All agricultural land that is put under any conservation status receives additional subsidies from the Brandenburg government. These subsidies are supposed to cover the extra costs that arise from the protection requirements for the farmers (ZALF et al. 2003: 54).

Brandenburg submitted its pSCIs and SPAs in three tranches in 1997/98, 2000 and 2005. The first tranche included 12 SPAs (7.6% of the Land area) and 90 pSCIs (1.3% of the Land area) all of which represented existing protected areas only. The second tranche consisted of 387 pSCIs (9% of the Land area). Most of these sites were also already part of some sort of conservation scheme, for example, 'conservation by contract' programmes in which farmers receive extra benefits for applying certain environment-friendly farming techniques. The third and final tranche was

---

[3] Unless otherwise indicated, this chapter is based on an inquiry carried out by Cand. Dipl.-Geogr. Kerstin Lehmann (Bonn) under the author's supervision.

submitted in 2005. In total, Brandenburg has notified 27 SPAs, covering 648,431 hectares or 22.0% of the Land area, and 620 SCIs, covering 332,842 hectares or 11.3% of the Land area. As some SPAs and SCIs overlap, in total 26% of Brandenburg's area has been designated as Natura 2000 sites (ZALF et al. 2003: 55; Zimmermann et al. 2000: 44; MLUV 2005a; 2005b).

The second and third notification tranches were prepared under extreme time pressure because the government did not adopt a pro-active approach in the nineties but rather had a strategy of reluctance and hesitation. As a consequence, the lists of pSCIs that had been prepared by the Environmental Agency could only be discussed to a very limited extent with relevant stakeholders, e.g. farming or conservation associations. There were almost no direct talks with landowners and land-users about specific conservation objectives and measures prior to submitting the pSCIs lists (ZALF et al. 2003: 55). Only municipalities and public agencies had the right to comment and submit objections.

The outcomes of this type of approach can hardly be surprising. Interviews with local stakeholders revealed that there is a very high degree of dissatisfaction about the whole Natura 2000 process. Many local stakeholders are not against Natura 2000 as a matter of principle, but complain that they had no information about the consequences of selecting an area as pSCI or SPA. Others stated that information was only delivered to them in a piecemeal fashion. Even a spokesman for the ministry admitted in a press interview that there had been 'communication problems' with NGOs and municipalities (Rath 2003). Stakeholders also criticized the formal public participation because there was not enough time for an in-depth assessment of the plans and because many objections had not been considered. Generally speaking, the flow of information and the opportunities for participation were judged as very insufficient. Consequently, many private land-users and landowners view Natura 2000 as a threat. They fear the conservation policy because it has been imposed on them and because it may bring economic disadvantages like decreased incomes and difficulties with further infrastructure developments and industrial investments.

18.3.3 Müritz National Park

The Müritz National Park is located in the southeast part of Mecklenburg-Western Pomerania, close to the border of Brandenburg. It consists of 32,000 hectares divided into two parts: one larger, western part on the shores of Lake Müritz, which is Germany's second largest lake after Lake Constance, and a smaller eastern part in an old-growth forest around the village of Serrahn. The national park features a richness of threatened species and biotopes, e.g. active raised bogs and populations of Osprey (*Pandion haliaetus*), Black Stork (*Ciconia nigra*) and Marsh Gentian (*Gentiana pneumonanthe*), to mention only a few (Leibenath 2001: 133). The Müritz National Park is entirely protected as an SPA according to the European Birds Directive and partly as SAC according to the Habitats Directive (NPAM 2004: 21).

The national park is administered by a national park authority with about 150 employees. There is also an advisory board ('Kuratorium'), consisting of high-ranking local politicians and representatives of public agencies, tourism organizations and other interest groups. The advisory board has to be involved in all major decisions concerning the management of the national park and any decisions that it takes must be unanimous (Leibenath 2001: 137).

Until 1989, most parts of today's national park were used as military training areas or as private hunting grounds by some of the German Democratic Republic's (GDR) socialist rulers. Officially, there were also about 15 nature conservation areas. The Müritz region was integrated into the GDR National Park Programme, which resulted in the designation of a number of large protected areas shortly before German reunification on 3 October 1990. The regulation on the establishment of the Müritz National Park was passed in September 1990 and published in the official journal of the GDR on 1 October 1990 under the old law (Ministerrat 1990). Due to the unique circumstances of the political turnaround, the designation process could be handled in a very straightforward way. There were some local initiatives in favour of a national park and the first plans were discussed with local representatives in round-table talks. However, the whole process did not last longer than nine months and was mainly promoted by some committed ecologists. The local population did not have any real chance to reflect thoroughly on the pros and cons of a national park in those turbulent times (Gaffert 1998: 19; Leibenath 2001: 135). This resulted in various forms of local resistance once the people had realized what it meant to live in or adjacent to a national park. Of course, the national park was also sometimes pushed into the role of scapegoat and made accountable for unemployment and other problems that were not necessarily caused by the national park.

The head of the national park authority Ulrich Meßner and his co-workers took a very open and cautious approach from the outset. They drew on the consequences from the confrontations that had occurred in other German national park regions, e.g. Schleswig Holstein's Wadden Sea National Park region, after detailed plans for the future development of those protected areas had been published by the respective conservation authorities. In avoiding the past mistakes of others, Meßner and his associates initiated a broad consultation and participation process when it came to preparing a management plan for the Müritz National Park, which was called the 'National Park Plan' (NPP).

A first meeting with public agencies and NGOs was held in 1993 when the preparation of the first draft of the NPP had just begun. The objective was to include regional concerns into the NPP right from the start (NPAM 1999: no page numbering). Later on, the national park authority, together with the advisory board, defined a procedure for adopting the NPP. The participation process started by introducing the draft version of the NPP in a number of public meetings. The park authority established a special information centre ('Bürgerbüro') and a telephone hotline. A flyer with information on the NPP and the adoption procedure was delivered to every household in the communities in or adjacent to the national park. The public could submit objections and comments to the NPP within eight months. In total, 905 proposals and comments were submitted, 53 of which were issued by communities or public agencies. The park authority and the advisory council formed a working group that included representatives of the counties, the municipalities and conservation authorities. The working group discussed each submission and decided whether and how it should be considered. At the end of the process, the NPP was modified in many ways. For instance, a vision for the national park was added and the chapter on how to interlink the national park with the region was revised. Everybody who had filed a submission received a written response about how his comment had been considered. All in all, the process took more than three years and was evaluated as fair and productive by the members of the working group (NPAM 2000: 19; NPAM 2001: 14; NPAM 2002: 21; NPAM 2003: 15). The three volumes of the NPP were published in 2003 and are available on the park's website (NPAM 2004).

18.4 Discussion and Conclusions

Particularly with regard to the Müritz case study, it can be said that there is an increasing emphasis on procedural legitimacy in processes of nature conservation policy formulation and implementation in Germany. The overview on public participation in the process of selecting pSCIs in the German Länder (Fig. 18.1) also made evident that more inclusive approaches have been adopted over time. On the other hand, the two case studies reveal major differences between the implementation of Natura 2000 and more traditional instruments like national parks. The European Habitats Directive and the related case law leave so little room for deliberation and consultation in the site selection process that local stakeholders very easily get the impression that their objections and comments will not be taken seriously and that the whole thing is nothing but a kind of show event. This is supported by findings from other Länder, which have employed more extensive forms of public participation than Brandenburg, e.g. Bavaria (Sauer et al. 2004: 45; Schreiber 2005).

Over the past few years, increasing efforts to make the Natura 2000 process more transparent can be observed. The conservation authorities of the Länder have improved and extended their information web sites. Most Länder governments not only publish maps and Standard Data Forms of SCIs but also detailed information on the designation process and the consequences for landowners and land-users (see e.g. BayStMinUGV 2005). However, there are also complaints about authorities that used outdated or incorrect ecological data, which nurtured doubts on the appropriateness of the delineated sites (Sauer et al. 2004: 31).

Throughput legitimacy depends on several factors but most importantly on accountability, meaning that decision-makers have to be identifiable and that it has to be possible to make them accountable. In the case of the Müritz National Park, the decision-makers were well-known figures on the local scene. In contrast, in the implementation of Natura 2000, it is often not at all clear who is actually in charge. Regional policy-makers contribute to this blurring of accountability by simply pointing to 'Brussels', which allegedly demands this or that without alternatives. This corresponds to the analysis of Scharpf (2004: 23), who identified three ways in which national governments can react to European policies that are not in line with national interests and preferences: firstly, governments can take on responsibility and defend the policy even if they, in fact, disagree with it; secondly, they can channel the protest towards the European institutions who are neither able to debate nor can be made politically accountable; thirdly, they can refuse compliance with the European policy and decline to implement it properly. In Germany, a clear bias towards the second and third options can be perceived.

By and large, there has been a shift from output legitimacy towards input and throughput legitimacy in German biodiversity policies. However, the implementation of Natura 2000 is not really a case in point. National parks, and some of the nature parks, are far better examples in this regard as has been shown in the Müritz case. The Kellerwald-Edersee National Park in Hesse would be another interesting case because it had been discussed in a decade-long process and was rejected in a regional referendum in 1997. However, after more years of discussions, consultations and information campaigns, a broad consensus in favour of a very strict form of conservation was reached throughout the entire region in 2004 (Harthun 1998 and 2004). As the Birds and Habitats Directives were based upon a merely ecological-technical approach towards conservation and prescribed a very tight time schedule, there was no room for similar forms of deliberative democracy. Although the European Commission is very much in favour of actively involving people who live in and depend on protected areas (Wallström 2000: 2), it did not play a major role in the shift towards procedural legitimacy in German biodiversity policies.

One could even argue that the provisions of the Habitats Directive reinforced German traditions of an elite approach to nature conservation, rooted in a deep sense of ecological mission among conservation officials and lobbyists. Conservationists who share the 'ecology first' paradigm tend to rely on 'professional management science' and to prefer hierarchical, top–down management styles. As Stoll-Kleemann (2001: 39) has precisely analysed, 'ecology first' conservationists depend on protecting specific species and habitats instead of implementing integrated land management plans, and they are unwilling to compromise on issues involving the fundamentals of ecological science. Instead, any local opposition is seen as a proof of the fact that other societal groups simply do not understand ecological concerns and just want to exploit nature. It is obvious that the Habitats Directive fits very nicely into such a worldview. By following this line of thought, one could conclude that the implementation of Natura 2000 stabilized anachronistic, output-oriented approaches to nature conservation that were already in decline in Germany and that otherwise would have disappeared much faster.

There are certainly trade-offs between the different types of legitimacy. If effectiveness is defined in terms of speed, then attempts to improve the input and throughput legitimacy of biodiversity policies can hardly be effective. The Müritz case study and also the example of the Kellerwald-Edersee National Park show that inclusive procedures can easily take a number of years. And, of course, there is a high risk that the original conservation goals get diluted in such a decision-making process or that the conservation effort completely fails. This can be counteracted by extensive information campaigns, which require a great deal of manpower, money, and again, time.

The implementation of European biodiversity policies has caused some severe coordination problems between different agents and policy levels in Germany, especially the procedural requirements of the Habitats Directive, which were not compatible with the existing distribution of power and competencies in the field of biodiversity policies. According to Knill (2003: 191–4), implementation problems are problems of institutional change. However, institutional change is only possible to the extent to which the fundamental identity of existing institutions is not put into question. This is what happened in the case of Natura 2000 in Germany because the Länder were no longer autonomous but had to coordinate and to consult on their conservation policies with national and European agencies. This is probably one of the reasons for the minimalist and ineffective implementation of the Birds and Habitats Directives in Germany (cf. Knill 2003: 215). In practice, there were several conflicts because the Länder had to implement Natura while the federal government was accountable to the European Commission. The Länder were responsible for the demanding and conflict-ridden process of selecting and designating sites. By contrast, the federal government was sued at the ECJ for insufficient implementation but hardly had any leverage against the Länder (SRU 2004: 122). This type of coordination problem might be most typical for EU member states with a decentralized, federal structure.

References

BayStMinUGV (=Bayerisches Staatsministerium für Umwelt, Gesundheit und Verbraucherschutz [Bavarian State Ministry for Environment, Public Health and Consumer Protection]) (2005). *Europäischer Biotopverbund Natura 2000: Wichtige Details für Grundstückseigentümer [European Biotope Network Natura 2000: Important Details for Landowners]*, available online: http://www.stmugv.bayern.de/de/natur/nat2000/detail.htm, 12.10.2005
Benz, A. (2000). 'Two Types of Multi-level Governance: Intergovernmental Relations in German and EU Regional Policy'. *Regional and Federal Studies* 10(3): 21–44
Berger, T. (2002). 'Unerhörte Lausitz. Regionaler Planungsverband will vorschnelle Meldung sächsischer Naturschutz-Gebiete an die EU verhindern' [Nasty Lusatia. Regional Planning Association wants to stop premature Proposal of Nature Conservation Areas to the EU]. *Sächsische Zeitung*, 11.02.2002, 6
BfN (=Bundesamt für Naturschutz [German Federal Conservation Agency]) (2005). *Status of Natura 2000 site proposals and implementation*, available online: http://www.bfn.de/en/03/030303.htm, 18.10.2005
Boedeltje, M., and Cornips J (2005). *Input and output legitimacy in interactive governance*, available online: http://www.demnetgov.ruc.dk/conference/int_conf/papers/MijkeBoedeltje_and_JuulCornips_paper.pdf, 09.09.2005
European Commission, GD Environment (2005). *Natura 2000 Barometer*, available online: http://europa.eu.int/comm/environment/nature/nature_conservation/useful_info/ barometer/barometer.htm, 18.10.2005
Frömmel, S. (2004). 'Die Rache der Hamster' [The Hamsters' Revenge]. *Die Zeit*, 22.04.2004, 63
Gaffert, P. (1998). Akzeptanzprobleme in Großschutzgebieten [Acceptance Problems in Large Protected Areas], in: K.H. Erdmann, N. Wiersbinski and H. Lange (eds.), *Zur gesellschaftlichen*

Akzeptanz von Naturschutzmaßnahmen – Materialienband [About the Societal Acceptance of Conservation Measures – Annex Volume], Bonn, Bundesamt für Naturschutz, 19–20

Getimis, P., and H. Heinelt (2004). *Leadership and community involvement in European Cities. Conditions of Success and/or Failure*, available online: http://www.uic.edu/cuppa/cityfutures/papers/webpapers/cityfuturespapers/session8_4/8_4leadershipcommunity.pdf, 16.09.2005

Harthun, M. (1998). 'Woran der Nationalpark Kellerwald vorerst scheiterte' [Why the National Park Kellerwald failed in the first place]. *Natur und Landschaft* 23(5): 223–227

Harthun, M. (2004). 'Der Nationalpark Kellerwald-Edersee. Hessens schwieriger Weg zum Buchenwald-Nationalpark' [The Kellerwald-Edersee national park. Hesse's difficult path towards a beech forest national park]. *Natur und Landschaft* 79(11): 486–493

Heiland, S. (1999). *Voraussetzungen erfolgreichen Naturschutzes: Individuelle und gesellschaftliche Bedingungen umweltgerechten Verhaltens, ihre Bedeutung für den Naturschutz und die Durchsetzbarkeit seiner Ziele [Preconditions of Successful Nature Conservation: Individual and Societal Factors for Environmental-friendly Behaviour, their Relevance for Nature Conservation and the Implementation of Conservation Goals]*, Landsberg am Lech, Ecomed

Job, H. (1996). 'Großschutzgebiete und ihre Akzeptanz bei Einheimischen: Das Beispiel der Nationalparke im Harz' [Large Protected Areas and their Acceptance among Local People. The Case of the Harz National Parks]. *Geographische Rundschau* 3/96: 159–165

Kehrein, A. (2002). 'Aktueller Stand und Perspektiven der Umsetzung von Natura 2000 in Deutschland' [State and Perspectives of the Implementation of Natura 2000 in Germany], *Natur und Landschaft* 77(1): 2–9

Knill, C. (2003). *Europäische Umweltpolitik. Steuerungsprobleme und Regulierungsmuster im Mehrebenensystem [European Environmental Policy: Governance Problems and Governance Patterns in the Multi-level System]*, Opladen, Leske+Budrich

Land Brandenburg (2005). *Das Land Brandenburg [The Federal Land Brandenburg]*, available online: http://www.brandenburg.de/cms/detail.php?id=65043&_siteid=20, 10.10.2005

Leibenath, M. (2001). *Entwicklung von Nationalparkregionen durch Regionalmarketing, untersucht am Beispiel der Müritzregion [Development of National Park Regions by Regional Marketing, based on a Case Study in the Müritz Region]*, Frankfurt, Peter Lang

Lucke, D. (1995). *Akzeptanz: Legitimität in der 'Abstimmungsgesellschaft' [Acceptance: Legitimacy in the 'Voting Society']*, Opladen, Leske+Budrich

MLUV (=Ministerium für Ländliche Entwicklung, Umwelt und Verbraucherschutz des Landes Brandenburg [Ministry for Rural Development, Environment and Consumer Protection of the State of Brandenburg]) (2005a). *Europäische Schutzgebiete in Brandenburg [European Conservation Areas in Brandenburg]*, available online: http://www.mlur.brandenburg.de/cms/detail.php?id=5lbm1.c.182175.de& _siteid=300, 12.10.2005

MLUV (=Ministerium für Ländliche Entwicklung, Umwelt und Verbraucherschutz des Landes Brandenburg [Ministry for Rural Development, Environment and Consumer Protection of the State of Brandenburg]) (2005b). *Aufbau von Natura 2000 [Establishing Natura 2000]*, available online: http://www.mluv.brandenburg.de/cms/detail.php?id=5lbm1.c.182564.de& _siteid=300, 12.10.2005

Möller, C. (1995). 'Strategien der Naturschutzverwaltung und Naturschutzpolitik: Vom Anwalt zum Mediator' [Strategies of Conservation Authorities and Conservation politics: From Advocator to Mediator]. *Politische Ökologie* 43: 28–32

NPAM (=Nationalparkamt Müritz [Administration of the Müritz National Park]) (1999), *Jahresbericht 1998 [Annual Report 1998]*, available online: http://www.nationalpark-mueritz.de/contents-de/1998.asp, 12.10.2005

NPAM (=Nationalparkamt Müritz [Administration of the Müritz National Park]) (2000), *Jahresbericht 1999 [Annual Report 1999]*, available online: http://www.nationalpark-mueritz.de/files/jber1999.doc, 12.10.2005

NPAM (=Nationalparkamt Müritz [Administration of the Müritz National Park]) (2001), *Jahresbericht 2000. Schwerpunktthema: Besuchermonitoring [Annual Report 2000. Thematic Focus: Visitor Monitoring]*, available online: http://www.nationalpark-mueritz.de/files/jber2000.doc, 12.10.2005

NPAM (=Nationalparkamt Müritz [Administration of the Müritz National Park]) (2002), *Jahresbericht 2001. Schwerpunktthema: Besucherinformation und -lenkung [Annual Report 2001. Thematic Focus: Interpretation and Visitor Management]*, available online: http://www.nationalpark-mueritz.de/files/jber2001.doc, 12.10.2005

NPAM (=Nationalparkamt Müritz [Administration of the Müritz National Park]) (2003), *Jahresbericht 2002 [Annual Report 2002]*, available online: http://www.nationalpark-mueritz.de/files/jber2002.zip, 12.10.2005

NPAM (=Nationalparkamt Müritz [Administration of the Müritz National Park]) (2004), *Nationalparkplan [National Park Development Concept]*, available online: http://www.nationalparkamt-mueritz.de/?id=10&file=nationalparkplan&lang=de, 12.10.2005

OJ (=*Official Journal of the European Union*), 24.07.2005, C 190, 17: Action brought on 8 April 2004 by Domäne Vorderriss, Rasso Freiherr von Cramer-Klett and Rechtlerverband Pfronten against the Commission of the European Communities (Case T-136/04) (2004/C 190/28)

Rath, I. (2003). 'Bringt FFH Aus für Oderangler?' [Will Natura 2000 be the Off for Anglers along the Oder River?]. *Märkische Oderzeitung* from 08.07.2003, available online: http://www.moz.de/index.php/Moz/Article/id/971, 12.10.2005

Rentsch, G. (1988). *Die Akzeptanz eines Schutzgebietes, untersucht am Beispiel der Einstellung der lokalen Bevölkerung zum Nationalpark Bayerischer Wald [Acceptance of a Protected Area, based on a Case Study on the attitudes of local people towards the Bavarian Forest National Park]*, Kallmünz/Regensburg, Michael Laßleben

Sauer, A., F. Luz, M. Suda and U. Weiland (2004). *Steigerung der Akzeptanz von FFH-Gebieten [Improving the Acceptance of Natura 2000 Sites]*, Bonn, Bundesamt für Naturschutz

Scharpf, F.W. (1998). *Interdependence and Democratic Legitimation*, available online: http://www.mpi-fg-koeln.mpg.de/pu/workpap/wp98-2/wp98-2.html, 09.09.2005

Scharpf, F.W. (2004). *Legitimationskonzepte jenseits des Nationalstaats [Concepts of Legitimacy beyond the Nation State]*, available online: http://www.mpi-fg-koeln.mpg.de/pu/workpap/wp04-6/wp04-6.html, 09.09.2005

Schreiber, R. (2005). *Umsetzung von Natura 2000 in Bayern – abschließende Gesamtmeldung 2004 [Implementation of Natura 2000 in Bavaria – Final List of Proposed Sites of Community Interest as of 2004]*, available online: http://www.bayern.de/lfu/tat_bericht/tb_200x/tb_2004/pdf/natura.pdf, 12.10.2005

SRU (=Rat von Sachverständigen für Umweltfragen [German National Environmental Advisory Council]) (2002). *Für eine Stärkung und Neuorientierung des Naturschutzes [Towards Strengthening and Reorientation of Nature Conservation]*, available online: http://www.umweltrat.de/ 02gutach/downlo02/sonderg/SG_Naturschutz_2002.pdf, 15.10.2005

SRU (=Rat von Sachverständigen für Umweltfragen [German National Environmental Advisory Council]) (2004). *Umweltgutachten 2004: Umweltpolitische Handlungsfähigkeit sichern [Environmental Policy Recommendations 2004: Ensuring Political Capacity for Environmental Action]*, available online: http://www.umweltrat.de/02gutach/downlo02/umweltg/UG_2004_lf.pdf, 21.06.2004

Stoll, S. (1999). *Akzeptanzprobleme bei der Ausweisung von Großschutzgebieten [Problems of Acceptance related to the Establishment of Large Protected Areas]*, Frankfurt, Peter Lang

Stoll-Kleemann, S. (2001). 'Reconciling Opposition to Protected Areas Management in Europe: The German Experience'. *Environment* 43(5): 32–44

Ssymank, A. (2005). 'Cross-border implementation and coherence of Natura 2000 in Germany' In: M. Leibenath, S. Rientjes, G. Lintz, C. Kolbe-Weber and U. Walz (eds.), *Crossing Borders: Natura 2000 in the Light of EU-Enlargement*, Tilburg and Budapest, European Centre for Nature Conservation, 25–39

Toogood, M., K. Gilbert and S. Rientjes (2004). *Biofact. Farmers and the Environment. Assessing the Factors that affect Farmers' Willingness and Ability to Cooperate with Biodiversity Policies*, Tilburg, ECNC

Umweltstiftung WWF Deutschland (1998). *WWF-Umfrage belegt: 70% der Deutschen wollen mehr Nationalpark [WWF opinion poll demonstrates: 70% of Germans want more National Parks]*, available online: http://www.wwf.de/c_bibliothek/c_presse/c_presse_newsarchiv/c_pm9805/c_presse_pm_980525.html, 05.06.1998

Wallström, M. (2000). Foreword, in: European Commission (ed.), *Managing Natura 2000 Sites. The provisions of Article 6 of the 'Habitats' Directive 92/43/EEC*, available online: http://europa.eu.int/comm/environment/nature/nature_conservation/eu_nature_legislation/specific_articles/art6/pdf/art6_en.pdf, 05.09.2005, 2

Wolf, K.D. (2002). *Civil Society and the Legitimacy of Governance Beyond the State – Conceptional Outlines and Empirical Explorations*, available online: http://www.isanet.org/noarchive/wolf.html, 16.09.2005

ZALF (=Leibniz-Zentrum für Agrarlandschafts- und Landnutzungsforschung e. V.), Fachgebiet Ressourcenökonomik der Humboldt-Universität zu Berlin & Institut für ländliche Strukturforschung (2003). *Halbzeitbewertung des Plans zur Entwicklung des ländlichen Raums gemäß VO (EG) Nr. 1257/1999 des Landes Brandenburg [Interim Evaluation of the Rural Development Plan according to the Council Regulation (EC) No. 1257/1999 of the Land Brandenburg]*, available online: http://www.mlur.brandenburg.de/cms/media.php/2317/halbzeit.pdf, 12.10.2005

Zimmermann, F., T. Schoknecht and A. Herrmann A (2000). 'Fachliche Kriterien für die Auswahl und Bewertung von FFH-Vorschlagsgebieten für das Fachkonzept NATURA 2000 in Brandenburg' [Technical Criteria for Selection and Assessment of Proposed Sites of Community Importance for NATURA 2000 in Brandenburg]. *Naturschutz und Landschaftspflege in Brandenburg* 9(2): 44–51

Legal Acts

Basic Law for the Federal Republic of Germany from 23 May 1949, (Federal Law Gazette, p. 1) (BGBl III 100-1), most recently amended by the amending law dated 26 July 2002 (BGBl I, p. 2863), available online: http://www.bundesregierung.de/static/pdf/GG_engl_Stand_26_07_02.pdf, 17.10.2005

Gesetz über den Naturschutz und die Landschaftspflege im Land Brandenburg (Brandenburgisches Naturschutzgesetz – BbgNatSchG) [Conservation and Landscape Planning Act for the Land Brandenburg], available online: http://www.mlur.brandenburg.de/cms/media.php/2318/bbgnatschg.pdf, 12.10.2005

Federal Nature Conservation Act of 25 March 2002 [Official Translation], available online: http://www.bmu.de/files/pdfs/allgemein/application/pdf/bundnatschugesetz_neu060204.pdf, 12.10.2005

Ministerrat (=Der Ministerrat der Deutschen Demokratischen Republik [Council of Ministers of the German Democratic Republic] & Amt des Ministerpräsidenten [Office of the Prime Minister]) (1990), Verordnung über die Festsetzung des Nationalparkes 'Müritz-Nationalpark' vom 12. September 1990 [Regulation on the Designation of the National Park 'Müritz National Park' of 12 September 1990], *Gesetzblatt der Deutschen Demokratischen Republik*, Sonderdruck Nr. 1468, 8

Hessisches Gesetz über Naturschutz und Landschaftspflege (Hessisches Naturschutzgesetz – HENatG) [Conservation and Landscape Planning Act for the German State of Hesse], available online: http://www.hmulv.hessen.de/imperia/md/content/internet/pdfs/umwelt/hessisches_gesetz__ber_naturschutz_und_landschaftspflege__hes.pdf; 12.10.2005

Treaty Establishing the European Community, Consolidated Version, *Official Journal of the European Communities* C 325, 24.12.2002, 33–159

Abbreviations

ECJ   European Court of Justice
GDR German Democratic Republic
NPP National park plan
pSCI Proposed site of Community importance
SCI   Site of Community importance
SPA   Special protected area

Chapter 19
Are Conflicts of Nature Distributive Conflicts?

Michiel Korthals

19.1 Introduction

Markus Leibenath has written a very clear chapter on the legitimacy processes with respect to biodiversity in Germany and the conflicts and impediments that Natura 2000 has encountered there. In particular, the specific character of the national traditions and institutions, the different tradeoffs between three different types of legitimacy and finally, the coordination problems between the different agents and polity levels have been illuminated in a very clear way.

At the very outset of his chapter, the author distinguishes between three different types of legitimacy and shows how a loss of input and throughput legitimacy occurred in the German implementation of Natura 2000 because of the policy's general top down approach (see Table 19.1).

In his discussion, Leibenath draws on two cases: Müritz National Park and the implementation of Natura 2000 in the land of Brandenburg. In the case of the Müritz National Park, Leibenath shows that the throughput and management process of the park received some attention by greater emphasis on participation and transparency. However, in the case of Natura 2000 in the land of Brandenburg, Leibenath explains that multi-level governance meant that the lower levels were confronted with an output-oriented policy from above, which caused a great deal of conflict, coordination problems, and delays because of protests by groups against the policy of exclusion. However, my main difficulty with the chapter concerns the claim that problems of biodiversity policy can be seen as problems of distributive justice or, in other words, of just and fair division of power. They can also be seen as *problems of misrepresentation*. Let me explain that.

Michiel Korthals
Department of Applied Philosophy, Wageningen University
E-mail: michiel.korthals@wur.nl

J. Keulartz and G. Leistra (eds.), *Legitimacy in European Nature Conservation Policy:*
Case Studies in Multilevel Governance, 251–255. © Springer 2008

Table 19.1 The three types of legitimacy and their transgressions

Legitimacy	Main features	Lack of type of legitimacy produces:
Input	Inclusion, agenda setting	Exclusion (policy as show event)
Throughput	Transparency, accountability	Secrecy, pressure (Brussels wants it)
Output	Objective needs, common interests	Issue or event rule, inefficiency, inefficacy

19.2 Conflicts over Concepts of Nature

According to Leibenath, the kinds of protest in the case of Natura 2000 in Brandenburg are protests against lack of power, which means protests against *unjust distribution of power*. However, protests and disagreements often have ambiguous meanings and are often difficult to decipher. In my view, protests and disagreements are often not only framed as conflicts of power, but also framed as conflicts over *deep-seated views on nature*, on the relationship between man and nature, and over the past neglect of these views. These competing concepts of nature cover cognitive aspects, but also normative (ethical and aesthetical) and social aspects (Korthals 2004). With respect to locations of nature, the question of access is a hot issue, which is an issue of power. However, the sheer value-impregnated character of biodiversity concepts makes conflicts over fundamental differences in views on nature often very emotional and very enduring. Most of the time, the stakeholders, like ecologists, neighbouring farmers, protest and civil society movements and policymakers cannot reach a consensual agreement on these views, and reasonable coexistence can be reached, at best. However, to reach a fair level of coexistence, respect has to be shown to the parties involved, and often parties do not feel respected because their views are not represented in the participation structures. Even if parties are represented in the participation structures, they may only feel partially represented or may feel compelled to formulate their view in terms of the dominating framework. In the case of biodiversity, it is striking how often parties use the same words but associate a different framework of meaning with the term (Eser 2003). It is even possible that parties feel so disrespected because of long neglect, that in the few cases that they can participate and are represented, they will only vent their anger and not act in a constructive way.

I can add a second point of criticism against Leibenath's view that these conflicts are only conflicts of unjust distribution of power. This point is connected to the first, but slightly different. In the cases of the designation of nature parks, it is often unclear for the stakeholders what the implications of certain nature arrangements will be. Consultations of stakeholders can have the meaning of constructive experiments to find out what the concrete character of the nature park should be. As such, these consultations are not about exchanging information and views of nature, but about discovering the main issues and problems. In short, the 'exchange of information' is not sufficient for legitimacy processes and for building trust, as is stated in article 2, paragraph 1, clause 15 of the Federal Nature Conservation Act: 'in the case of measures planned for nature conservation and landscape management, the early exchange of information with the parties concerned and interested segments of the general public shall be safeguarded' (cited by Leibenath in section 2.1).

Giving everyone the power to exchange information is not the same as allowing everyone to participate in agenda setting or coming to the best solution. In working against the misdistribution of power, one can give everyone the same right of participation, but this does not say anything about the type of values and value laden opinions that are allowed in the arena and the equal respect these value laden opinions get. It still could be that the dominant view of nature rules, without thoroughly being discussed, and that others are demeaned and debased or just crushed.

19.3 Misdistribution Versus Misrepresentation

In general, I think it is necessary in deciphering conflicts of nature to cover not only misdistribution of power and coordination problems between agencies on the different levels, but also to cover the way problems are framed and the possible misrepresentation of broad views on nature that are only partially tolerated and accepted or sometimes not even heard at all (see Table 19.2).

Misrepresentation of views has several aspects. As previously said, what is misrepresented or not represented at all are opinions, views, mostly paradigms or frameworks that determine one's basic attitude towards nature and select what information counts and what actions are seen as important and what not. These visions frame the relevant features of pieces of information and what they determine as not relevant will be left out of the discussion. Misrepresentation concerns the repression of fundamental concepts of nature such as rural or agricultural nature and can bring about severe emotional reactions because misrepresentation goes deep to the heart of a group's identity. A vision of nature determines the way a group sees itself and is seen by others. If a group feels misrepresented with respect to its vision of nature, it feels disrespected and neglected, even held in contempt (see Ricoeur 2005). Ideally, fair and just representation would give everyone the opportunity to make his or her vision of nature public. This is exactly what fair representation means: that every relevant view is represented.

Disrespecting certain people is very often more functional in establishing a power base than outright disempowerment of people. In the nature discussion, Leibenath describes one type of vision of nature that is pushed and accompanied by ridiculing others and demeaning them as unscientific, non-ecological, destructive for nature or merely traditional. Emphasizing the urgency of ecological measures, the visions of others are downplayed as merely delaying tactics. Whereas in distributive conflicts the main bone of contention is covered by the question of who decides, in conflicts of misrepresentation the main issue is what vision dominates and what type of information is allowed. Because the underlying vision of nature determines the selection of information, the debate is very often sparked by questioning certain types of information as relevant or not relevant. In Germany, as in other Western nations, the vocabulary on nature that has framed most of the discussion in the last twenty years is that of ecology.

Table 19.2 Combining seven aspects of misdistribution and misrepresentation

	Misdistribution of power	Misrepresentation of nature/biodiversity
Social Evil	Misdistribution (no right of access), exclusion	Repression of certain nature concepts, i.e. important locations, access and species (no right of voice)
Ethical Evil	Injustice (no equal distribution)	Disrespect of certain nature conceptions (against sovereignty, voice)
Ethical Emotions	Feeling a lack of power	Feeling of humiliation
Social Abuse	Nature as power instrument: blackmailing with nature (birds etc.)	Nature (Locations, birds etc.) as instrument of repression: disrespect
Moral Legitimation (Ideology) of Exclusion	Not included do not know what nature is	Disrespected nature concept: nature is still exploited etc.
Power Object	Farmers, users of the land, owners	Users of the land, other stakeholders (citizens)
Governmental Action	Nature as instrument against inner opposition, against Länder	Politics against unwilling minorities

The two main ethical perspectives on injustice – misdistribution and misrepresentation – often, though not always, come together. We can create a two-by-two matrix that provides us with four possibilities. The classic worst case is that of the proletariat in the nineteenth century as analyzed by Marx and Engels. The labour class at that time did not have enough income, wealth or power, but its views were also not taken seriously. The labour class was simply seen as mean, stupid and uneducated. The recent struggles of nature have mostly resulted in the victory of groups that do not differ from the defeated groups in resources but in ideology. These recent struggles illuminate the other extreme, where just distribution goes together with misrepresentation (see Table 19.3).

Table 19.3 Four relations between (mis)distribution and (mis)representation

	Representation of visions	Misrepresentation of visions
Just Distribution (Fair Division of Power)	Represented and equal (utopia): there are no underprivileged groups	Equal but not represented: Europe and GM food; struggles of nature
Misdistribution (Differences in Power)	Not equal but represented: minority dominates culture	No power and not represented: classic case of deprivileged groups

19.4 Conclusion

In conflicts of nature, it is necessary to take into account not only the distribution of power and the power differences between social levels, but also the way that visions of nature are represented at these levels and the exclusion mechanisms that are used in trying to wipe out some and make others dominant. Misrepresentation is as important in debates on nature as is misdistribution. As a consequence, giving information on and being transparent about the decision procedures are not the only political requirements. Because visions of nature determine what counts as relevant information, they should also be explicitly put on the political agenda, not in misrepresented form, but in fair recognition of their main features.

References

Eser, U. (2003). 'Der Wert der Vielfalt: "Biodiversität" zwischen Wissenschaft, Politik und Ethik. In M. Bobbert, M. Düwell and K. Jax, eds. *Umwelt – Ethik – Recht*. Tübingen: Francke, pp. 160–181.
Korthals, M. (2004). *Before Dinner*. Dordrecht: Springer.
Ricoeur, P. (2005). *The Course of Recognition*. Cambridge: Harvard University Press.

Part V
Conclusions

Chapter 20
European Union Environmental Policy and Natura 2000

From Adoption to Revision

Rüdiger K. W. Wurzel

20.1 Introduction: Natura 2000 Within the Wider EU Environmental Policy Context

Natura 2000 constitutes the cornerstone of the European Union's (EU) nature protection policy and its more recent biodiversity strategy (CEC 2006). It developed out of the 1979 wild birds directive (79/409) and 1992 habitats directive (92/43) while funding for Natura 2000 measures has been provided by the LIFE nature fund. In order to understand better the implementation difficulties which occurred with Natura 2000 measures on the member state level and the problems that have dogged its revision process on the EU level, it is necessary to assess the adoption process while analysing Natura 2000 within a wider EU environmental policy context. As Paul Sabatier (1993) has argued convincingly, it is possible to assess the full impact (or lack of impact!) of public policy measures only by taking into account a period of time which stretches for more than a decade.

20.2 Phases of EU Environmental Policy

Broadly speaking, the following four EU environmental policy phases can be identified (see also Hildebrand 1992):

(1) 1958–72: Incidental phase;
(2) 1972–87: Adolescent phase;
(3) 1987–92: Mature phase;
(4) 1992-: Defensive and reorientation phase.

Rüdiger K. W. Wurzel
Director, Centre of European Union Studies (CEUS), University of Hull, UK
E-mail: R.K.Wurzel@hull.ac.uk

J. Keulartz and G. Leistra (eds.), *Legitimacy in European Nature Conservation Policy: Case Studies in Multilevel Governance*, 259–282. © Springer 2008

20.2.1 Incidental Phase (1958–72)

Prior to 1972, EU environmental policy measures were adopted in an incidental manner and primarily concerned with the prevention of trade barriers (Bungarten 1978; Rehbinder and Stewart 1985). Because environmental issues were only of a very low political salience in Europe during the 1950s, environmental policy (including nature protection) was not explicitly identified as an area for common policy measures in the founding Treaties (i.e. the 1952 European Coal and Steel Community (ECSC) Treaty, 1957 European Economic Community (EEC) Treaty and 1957 European Atomic Energy (EURATOM) Treaty) out of which the EU has grown. After World War II, economic recovery and political reconciliation were clearly given a significantly higher priority than nature protection.

The first incidental common 'environmental' policy measures were aimed mainly at the prevention of barriers to trade rather than the protection of the environment because the EU's primary goal was the creation of an internal market. A typical example of an 'environmental' policy measure during the incidental phase is the directive on the classification, packaging and labelling of dangerous substances (67/548). Its main aim was to facilitate the (safe) trade of dangerous substances within the internal market by harmonising member states' rules on the EU level. Unsurprisingly not a single EU nature protection measure was adopted during the incidental phase.

20.2.2 Adolescent Phase (1972–87)

The beginning of the adolescent phase was marked by the Heads of State and Governments of the member states who gathered for a European Council[1] meeting in Paris in October 1972. The 1972 European Council adopted the following declaration:

> [E]conomic expansion, which is not an end in itself, must...emerge in an improved quality as well as improved standard of life. In the European spirit special attention will be paid to non-material values and wealth and to the protection of the environment so that progress shall serve mankind (EC Bulletin 1972: 10 and 15–16).

This somewhat lofty statement was adopted only a few months after the United Nations (UN) had staged it first major international conference on the environment in Stockholm in June 1972. It reflected both the rise of postmaterialist values in Europe (Inglehart 1971) as well as the desire of the member states' political leaders to lend a human face to the European integration project which, up to then, had focused largely on the internal market.

At the 1972 Paris summit, the Heads of State and Governments asked the Commission to draft the First Environmental Action Programme (EAP) which was

---

[1] The summits of the Heads of State and Governments of the member states were formally called European Council meetings only from the mid-1970s onwards. Denmark, Ireland and the United Kingdom also attended the 1972 Paris summit, joining the EU only in January 1973.

adopted in 1973. The First EAP outlined the main policy principles and indicated the EU's medium term environmental priorities. However, although the programme was legally non-binding it provided the Commission with additional legitimacy for proposals for EU environmental policy (including nature protection) measures. The Commission formally has to put forward proposals for all legally binding EU policy measures which are then adopted by the Council of Ministers and the European Parliament (EP).

At a meeting in 1972, the ministers responsible for environmental policy – not all member states had yet set up an Environment Ministry – agreed in the Council of Ministers that proposals for EU environmental policy measures could be based on (1) the EAPs; (2) the so-called Information and Standstill Agreement; and, (3) existing Treaty provisions which included the internal market provision (article 94, formerly article 100), that allows for the harmonisation of nationally different standards affecting the functioning of the internal market, and the so-called implied powers doctrine (article 308, formerly article 235). These three main sources of legitimacy for common environmental policy measures were identified by the ministers responsible for environmental policy because the Heads of States and Governments had failed to agree on Treaty amendments that would have introduced explicit environmental policy competences for the EU (Bungarten 1978; Wurzel 1993). Explicit environmental Treaty provisions arguably would have paved the way for a wider range of nature protection measures in the adolescent phase during which the nascent EU nature protection policy was strongly influenced by international developments.

The internal market provision clearly became the main driver for most EU environmental policy measures until the 1987 Single European Act (SEA) introduced explicit environmental policy Treaty provisions. This explains how the EU managed to adopt more than 100 legally binding common environmental policy measures before 1987 (Haigh 2005; Weale 1996; Weale et al. 2000; Wurzel 2002). However, the internal market provision proved largely ineffective for the adoption of EU nature protection measures because nationally different nature protection measures do not usually have a direct impact on the functioning of the internal market. There was therefore normally no need to demand the harmonisation of these national measures on the grounds that they constituted barriers to trade.

The implied powers doctrine, which is typical for federal systems, became an important legal base for the steady trickle of EU nature protection measures that set in already prior to the 1987 SEA. The implied powers doctrine was, however, frequently contested by the French and British governments in particular because they were concerned about the expansion of EU powers into policy areas (such as environmental policy) which were not explicitly covered by the Treaties. However, the Commission nevertheless based some of its proposals for EU environmental legislation, including the 1979 wild birds directive on the implied powers doctrine (see CEC 1992: 2). The introduction of explicit Treaty competences of environmental policy by the 1987 SEA rendered the implied powers doctrine largely irrelevant as a legal base for EU nature protection measures.

Another commonly used source for the legitimisation of common environmental policy measures was the Information and Standstill Agreement which had originally been adopted in 1969 in order to facilitate the creation of an internal market. It required member governments to inform the Commission about draft national legislation with a potential impact on the internal market. Member states were required to put on hold their draft national legislation for up to one year in order to allow the Commission time to propose EU legislation instead. In 1973, the Information and Standstill agreement was amended in order to cover explicitly also draft environmental legislation with a potential impact on the internal market. However, only very few national (draft) nature protection laws actually had a direct impact on the internal market. The Information and Standstill Agreement therefore also failed to provide a real impetus for the EU's gradually emerging nature protection policy.

The 1973 First EAP, although it was not legally binding, constituted a considerably more important source of justification for the EU's early nature protection measures. It contained a chapter entitled 'protection of the natural environment' (CEC 1973: 38–40) which proposed the adoption of measures for the 'protection of birds and certain other animal species' while pointing out that '[h]undreds of millions of migratory birds and songbirds are captured and killed in Europe every year provoking worldwide protests against the countries which allow the trapping of birds' (CEC 1973: 40). Nigel Haigh (2005: 9.2–1) has therefore concluded that the wild birds directive 'arose out of public disquiet at the annual slaughter of migratory birds that was common in southern Europe, but goes further in providing a general system of protection for all species of wild birds found in Europe'. The 1973 First EAP proposed 31 December 1974 as deadline for the adoption of EU protection measures for wild birds. Recognising the importance of international developments as a driver for EU nature protection it also suggested (in addition to EU measures) 'joint action by the Member States and the Council of Europe and other international organizations' (CEC 1973: 40).

However, following protracted negotiations in the Council of Ministers the wild birds directive was adopted only in 1979 (see Table 20.1). France and Italy in particular had held up the Council negotiations by insisting on derogations for the prohibition of hunting certain bird species (Haigh 2005: 9.2–4). It was nevertheless adopted unanimously – as all EU environmental legislation had to be prior to 1987 – despite a relatively weak legal base (i.e. implied powers doctrine) and considerable unease in some member states (mainly in France and Italy) about certain obligations of the wild birds directive.

Nigel Haigh (2005: 9.9–6) has rightly pointed out that '[n]ature conservation was not entirely overlooked in the early years of the Community's environmental policy but it was secondary to the primary concern with the control of pollution'.

20.2.3 Mature Phase: 1987–92

The beginning of the mature phase is marked by the coming into force of the SEA in 1987. The SEA introduced an environmental chapter into the Treaty, stipulated qualified majority voting (QMV) (for the revised internal market provision which explicitly mentioned environmental policy) and introduced the co-operation proce-

Table 20.1 Natura 2000 measures within the wider EU environmental policy context

Time period	General descriptor and specific Natura 2000 measures
1958–72	Incidental phase: • No EU nature protection measures adopted
1972–87	Adolescent phase: • 1979 Birds directive (79/409)
1987–92	Mature phase: • 1992 Habitats directive (92/43)
1992-	Defensive and reorientation phase: • Gradual recognition of the scale of the implementation deficit • 1997 Commission decision concerning information for Natura 2002 sites • 1997 Council directive amendingAnnexes I & II: Cormorant removed from Annex I • LIFE and LIFE + funding for nature protection and biodiversity measures

dure which granted the EP more powers. The adoption of EU environmental laws continued to rise gradually during the mature phase although it did not jump suddenly as can be seen from Fig. 20.1.

The mature phase of EU environmental policy is very much a continuation of the adolescent phase. There were, however, also at least three new features which had an important impact on the EU's environmental policy in general and its nature protection policy in particular. First, when the SEA came into force in 1987, environmental

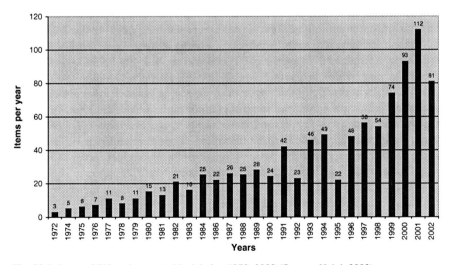

Fig. 20.1 Items of EU environmental legislation 1972–2002 (Source: Haigh 2003)

policy measures (including nature protection measures which had only a tenuous link to the internal market) could be based on explicit Treaty provisions (without the need of making recourse to the contested implied powers doctrine). Second, the EP, which is widely seen as the 'greenest' of all EU institutions, gained a bigger say in the (environmental) policy-making process when the 1987 SEA introduced the co-operation procedure. Third, up to the mature phase EU environmental policy measures had consisted almost exclusively of traditional 'command-and-control' regulations. They were largely adopted in top down fashion by EU policy makers on the supranational level but had to be implemented on the member state level by national and/or local actors. The 1987 Third EAP (CEC 1987) was the first action programme which flagged up the use of market instruments (such as eco-taxes) and voluntary agreements as possible instruments for EU environmental policy. However, as can be seen from Fig. 20.1, it did not slow down the adoption rate of EU traditional ('command-and-control') environmental legislation.

Fourth, while the number of common environmental laws continued to rise during the mature phase, there was a gradual recognition that many EU policy measures were not properly implemented on the member state level. After pressure from the EP, the Commission started to take seriously the implementation deficit in the environmental policy field. Environmental groups often acted as whistleblowers drawing to the Commission's attention failures by member states to implement EU environmental legislation (Mazey and Richardson 1993). From the late 1980's onwards, the Commission, which acts as the guardian of the Treaties, took to the European Court of Justice (ECJ) a growing number of member states which it suspected of failing to implement EU environmental legislation. Subsequently, almost all member states found themselves in front of the ECJ for being in breach with the 1979 wild birds directive (and later also the 1992 habitats directive). The ECJ often backed the Commission with its usually pro-integrationist judgements.

Under Environmental Commissioner Ripa di Meana (1989–92), the Commission repeatedly risked public confrontations with member governments (Wurzel 2002: 66). However, although Ripa di Meana was praised by environmental NGOs, his naming and shaming strategy caused ill-feeling among some member governments who complained about the EU's interference with the nooks and crannies of member state affairs as the then British Foreign Minister, Douglas Hurd, put it. Ripa di Meana's strategy would have been untenable without the unprecedented upsurge in public environmental awareness in the early 1990s. However, the Commission's confrontational implementation strategy triggered a fairly hostile reaction from (some) member states when the European integration project was threatened by the 'No' vote in the first Danish referendum on the Maastricht Treaty which had been signed by the Heads of State and Governments in 1992.

20.2.4 1992-: Defensive and Reorientation Phase

1992 constituted a turning point in the history of EU environmental policy. The first half of 1992 was characterised by great optimism and environmental activism. The preparations for the UN Earth summit in Rio in June 1992 were in full swing and the Commission was hopeful that its proposal for a common carbon dioxide/energy tax would be adopted on time for the summit. On the surface, the EU looked able to present itself at the 1992 UN Rio summit as a model for regional environmental co-operation. However, the carbon dioxide/energy tax was never adopted and the Danish 'No' in June 1992 triggered a debate about the principle of subsidiarity which pushed EU environmental policy onto the defensive and triggered a reorientation of EU environmental policy.

In June 1993, a 'hit list' was drawn up by the British and French governments which proposed the repatriation of more than 100 EU laws including 24 environmental laws (Wurzel 2002). Although none of the EU environmental laws listed were later scrapped, the Commission's Directorate General (DG) for Environment started to propose fewer detailed technical regulations while relying more strongly on flexible framework directives, cost-effectiveness appraisals and procedural measures (such as environmental impact assessments).

The habitats directive (92/43), which 'extends many of the protection mechanisms established for birds in Directive 79/409 to other species and habitat types' (Haigh 2005: 9.9–1), was adopted at the beginning of the defensive and reorientation phase although it had been proposed by the Commission already in 1988. It constitutes very much a classic top down measure in terms of the adoption process and a traditional ('command-and-control') regulatory tool in terms of the implementation process. As has already been pointed out in the contributions to this book, the habitats directive (92/43) aims to bring about 'favourable conservation status' for both natural habitats and wild species. In 1982, decision 82/72 had already implemented the Convention on the conservation of European Wildlife and natural habitats (Bern Convention) which shows the importance of the international level on EU nature protection policy during the adolescent phase.

However, the habitats directive 92/43 went further than decision 82/72 by introducing a European ecological network Natura 2000, for special areas of conservation (the details of which have been explained in the previous chapters of this book). The Natura 2000 network includes areas which contain the natural habitats given in Annex I and the habitats of the species given in Annex II (CEC 1996: XI). The Natura 2000 network also includes the special protection zones which have been set up by member states under the wild birds directive (79/409).

Importantly, the implementation of the habitats directive (92/43) has been co-financed by the EU's so-called structural funds (i.e. regional policy funds) and, more importantly, a specific financial instrument – the LIFE nature fund. As Nigel Haigh (2005: 9.9–5) has pointed out '[i]t has been accepted that the Polluter Pays Principle has limited application to nature conservation and that an excessive financial burden will be imposed on some member states because of the uneven distribution of habitats and species'. Spain in particular lobbied hard for EU funding of Natura 2000 measures (Haigh 2005: 9.9; see chapter 6 by Susana Aguilar Fernández). Some member states, especially those who are net contributors to the EU's budget, opposed co-financing on the grounds that it would constitute a breach of the polluter pays principle and set a dangerous precedent (Haigh 2005: 9.9–8). However, they were finally swayed by the argument that without co-financing the financial burden would otherwise be grossly disproportionate for some of the poorer Southern member states in particular. Since 1999, the Commission has repeatedly warned that very poor implementation of the wild birds directive (79/409) and the habitats directive (92/43) might put member states' entitlement to regional funding under the Structural Funds (Haigh 2005: 9.9–15). However, the existing amount of EU funding available for the implementation of Natura 2000 measures is inadequate.

A Commission expert working group initially estimated the cost for implementing Natura 2000 measures amounts to between at least €3.4 and 5.7 billion per year between 2004–2013 (Haigh 2005: 9.9.17). However, this is a very conservative estimate. A committee of the German states *(Länder)* calculated that the implementation cost for Natura 2000 in Germany alone will amount to approximately €4.9 billion between 2003–2012 (SRU 2004: 127). More recent estimates have put the cost of the Natura 2000 network of protected areas at €6 billion per year for the EU made up of 25 member states (EU-25) *(EEPL* 26 May 2006). However, in 2006, Environmental Commissioner Stavros Dimas has defended the EU's nature protection policy and biodiversity strategy by arguing that '[d]epleted fish stock, declining soil fertility, the collapse of pollinator species, the damage done by invasive species and the loss of potentially valuable genetic resources all reduce economic well-being' *(ENDS Europe Daily* 30 May 2006). Dimas also warned that 'continued biodiversity loss would seriously threaten "ecosystem services", including foods, fibres and medicines, worth an estimated €26 trillion per year' *(ENDS Europe Daily* 22 May 2006).

Already the 1977 Second EAP (CEC 1977) suggested the need for EU measures to protect wildlife and biotopes while the 1987 Fourth EAP (CEC 1987) stated that 'the time is ripe for the Community and the Member States to make a major new thrust in the field of nature protection' (cited in Haigh 2005: 9.9–7). However, the habitats directive (92/43) was adopted only in 1992. It enshrined many of the obligations of the above mentioned Bern convention but it took 'it a step further, particularly by seeking the protection of certain types of habitat (often referred to by their scientific terms as biotopes) for their own sake rather than because they harbour valued species' (Haigh 2005: 9.9–7). The above quoted declaration by the 1972 European Council, which stated that '[i]n the European spirit special attention will be paid to non-material values and wealth and to the protection of the environment so that progress shall serve mankind' *(EC Bulletin* 1972: 10 and 15–16) seems to resonate in Haigh's analysis of some of the objectives of the habitats directive. However, one of the main justifications for the habitats directive (92/43) was to improve 'the effectiveness and coherence' of the Bern convention which suffered from a poor implementation record by its signatory states (Haigh 2005: 9.9–8).

Nigel Haigh (2005: 9.9–7) has argued that Stanley Johnson, a British national who worked as a senior official for DG Environment, was 'closely associated with the drafting of both the fourth Environmental Action Programme and the habitats Directive'. However, some member states (such as Euro-sceptical Denmark and the United Kingdom) were concerned about the EU's expansion of powers into new areas such as habitats protection (Haigh 2005: 9.9–7). Council negotiations of the habitats directive therefore made real progress only after the 1987 SEA introduced explicit environmental competences and qualified majority voting (QMV) to the Treaty.

The text of the draft directive 'changed significantly between 1988 and December 1991 when political agreement. . .was reached under the Dutch Presidency' (Haigh 2005: 9.9–8). The habitats directive (92/43) has therefore been perceived very much as a 'Dutch directive' by many officials in the Netherlands who felt that it came close to existing domestic nature protection legislation (interviews with former Dutch officials in 2005). However, although the Netherlands significantly influenced the final text of the habitats directive it – like all member states – later had considerable difficulties in implementing this piece of EU environmental legislation (interviews with former Dutch officials in 2005; see also chapter 2 by Leistra, Keulartz and Engelen).

There are three main reasons for the implementation difficulties in the Netherlands and other member states. First, the habitats directive – like all pieces of EU (environmental) legislation – constitutes a compromise which accommodates the core demands of veto actors such as member state governments and the EP (Weale 1996; Weale et al. 2000; Wurzel 2002). The Netherlands was therefore not able simply to 'upload' to the EU level its national nature protection legislation. The habitats directive contains provisions which were 'alien' to the Dutch nature protection approach and thus created problems in terms of the 'goodness of fit' (Börzel 2002; Knill 2001; Knill and Lenschow 2000). This helps to explain why it is not only the new member states such as Poland (see chapter 10 by Tadeusz Chmielewski and Jaroslaw Krogulec), which joined the EU only after the Natura 2000 measures had been adopted, but also the old member states which have had serious implementation difficulties. Second, member governments often underestimate the implementation cost when they agree to environmental legislation on the EU level. During the implementation phase it became clear that the cost for the implementation of Natura 2000 measures will be much higher than initially estimated by the Commission. However, as with most environmental policy measures the short to medium term costs are more easily identifiable than the medium to long-term environmental benefits. Moreover, the short to medium term costs often fall on a limited number of actors who find it relatively easy to organise collective resistance and/or demand side-payments in the form of subsidies for costly implementation measures while the medium to long term benefits (in terms of biological diversity and quality of life) are often widely dispersed and difficult to estimate in monetary terms so that collective action is significantly more difficult to organise in favour of environmental legislation (see, for example, Weale 1992). Third, the 1992 habitats directive was adopted very much in a top down fashion with relatively little consultation of stakeholders and policy actors on the ground who have to implement the Natura 2000 measures.

The defensive and reorientation phase also saw the adoption of several, at times contradictory, meta strategies some of which have a significant impact on the EU's nature protection policy and biodiversity strategy. The most important recent meta strategy is the Lisbon strategy which was adopted in 2000 with the aim of making the EU 'the most dynamic and competitive knowledge-based economy in the world' by 2010. However, the Lisbon strategy initially failed to mention environmental policy which was inserted only retrospectively under the Swedish EU Presidency at the Gothenburg summit in 2001. The EU's 2001 sustainable development strategy gives equal weight to economic, environmental and social concerns (CEC 2001). However, in 2005, Commission President Manuel Barroso made it clear that the Commission would pay more attention to economic policy rather than to environmental or social concerns by likening his role as Commission President to a father who has to pay more attention to a sick child than the healthy children while implying that Europe's economy has been sick for some time (Wurzel 2007). However, only one year later, the Environmental Commissioner, Stavros Dimas, claimed that the EU is determined 'to integrate biodiversity concerns into other policy areas, including the Lisbon agenda, rural development and agriculture, and regional development' (*ENDS Europe Daily* 22 May 2006) although no practical integration measures have since been taken. It is therefore not clear how the EU's Lisbon strategy can be synchronised with Natura 2000 implementation and revision efforts and the aim of halting the loss of biodiversity by 2010 as put forward in the biodiversity strategy which the Commission published in 2006 (CEC 2006).

Better regulation and implementation are themes which have gained greatly in importance on the EU level since early 2000s. The Commission's *White Paper on European Governance* (CEC 2001) called for the opening up of the EU policy-making process in order to get more stakeholders involved in shaping and delivering EU policies. The Governance White Paper also promoted greater openness, accountability and responsibility. Better regulation is one of the overarching themes of the Governance White Paper which proposed, amongst others, greater use of different policy instruments including framework directives (which regulate only the essential issue while leaving the detail up to member states) and co-regulation (which combines traditional regulation with [voluntary] actions taken by the actors most concerned while drawing on their practical expertise in the hope that this might result in wider ownership of the policies and thus better implementation). Non-regulatory instruments such as the open method of co-ordination (OMC) have also been advocated in the Governance White Paper. The OMC, which, so far, has been used mainly for EU employment and social policies, essentially consists of the following three-step process: (1) encourage co-operation between member states; (2) exchange best practice (i.e. benchmarking); and, (3) agree on common targets and guidelines. However, recent research on 'new' environmental policy instruments (NEPIs) has found that despite widespread demands for more 'non-regulatory' instruments and wider participation of stakeholders in recent years, traditional ('command-and-control') regulation still dominates both EU and member state environmental policies (Jordan et al. 2003; 2005; 2006). However, before concluding that there is little empirical evidence of the alleged shift from (top down) *government* towards (bottom up) *governance* which, according to some observers, has taken place in most highly developed liberal democracies in recent years (see also chapter 1 by Engelen, Leistra, and Keulartz), it is necessary to assess briefly the government-governance debate before returning to the main themes flagged up in the previous chapters of this volume.

20.3 From Government Towards Governance?

Political scientists have traditionally used synonymously the terms government and governance (e.g. Finer 1970). However, in recent years it has become increasingly common to differentiate analytically between governance and government (Jordan et al. 2003; 2005; 2006; see also the introductory chapter by Engelen, Leistra and Keulartz). Much of the recent governance literature argues that government is essentially about the use of 'command-and-control' regulation while governance relies on less interfering bottom up, soft and/or participatory policy instruments and strategies. In essence governance advocates claim that governance is characterised by an increased level of direct involvement of stakeholders in the policy-making process which, in extreme cases, may lead to 'self-steering' of societal actors through a process of self-regulation. The governance perspective propagates that central government is playing an increasingly less important role thus leading to a reduction of ('command-and-control') regulations and a decline in top down implementation strategies (for a more detailed analysis see Jordan et al. 2003; 2005; 2006; see also the introductory chapter by Engelen, Leistra and Keulartz). However, as Fritz Scharpf (e.g. 1993) has pointed out, most voluntary agreements and/or self-regulatory measures are adopted by societal actors in 'the shadow of the law'. Without the threat of government regulation many of them would not have been agreed.

Osborne and Gaebler (1992) have distinguished between 'steering' and 'rowing' when arguing that some policy networks 'steer' by setting policy goals while others 'row' by delivering the policy goals (see also Jordan et al. 2005). Since the 1990s, there has been a shift in EU environmental policy away from an almost exclusive focus on 'output legitimacy' (i.e. the delivery of environmental policy goals) towards greater attention for 'input legitimacy' (e.g. wider participation of stakeholders) and 'through put legitimacy' (i.e. procedural measures such as environmental impact assessment) (for a detailed analysis of the analytical distinction between the different types of legitimacy see the introductory chapter by Engelen, Leistra and Keulartz). However, this shift in emphasis has been a contested one. For example, member states such as Germany and Austria, which have traditionally relied heavily on detailed environmental standards (i.e. 'command-and-control' regulations) derived from the best available technology (BAT) principle have opposed the increased use of procedural measures in EU environmental policy (Wurzel 2002; 2003; 2004). In Germany the greater emphasis on procedural measures (i.e. input legitimacy) has variously been attributed to 'Anglo-American' (Kloepfer 1998: 614), 'British' (Héritier et al. 1996; Knill 2001) or 'Anglo-Scandinavian' (Baacke 2000) influence. Most German environmental policy makers have perceived it as a threat to ambitious detailed environmental standards which focus mainly on output legitimacy (Wurzel 2003; 2004).

Another recent trend in EU environmental policy also has important consequences for the common nature protection policy. Since the 1990s, attempts have been made to integrate better environmental requirements into all sectoral policies (such as the common agricultural policy (CAP) and transport policy). However, the so-called Cardiff strategy, which was initiated by the British government during its EU Presidency in 1992 has failed (Information gathered from interviews by the author with British and Council General Secretariat officials in 2006). Its aim was that all technical Council of Ministers formations (e.g. the Agricultural Council and the Transport Council) should take more seriously environmental considerations (including nature protection requirements). The integration of environmental requirements into other policy sectors therefore has to be conducted primarily through the EU's sustainable development strategy.

When trying to improve the implementation of Natura 2000 measures the EU has traditionally relied very heavily on top down 'government strategies'. Bottom up 'governance strategies' (such as the wider consultation of stakeholders) are more recent measures whose use has not (yet) been widespread in nature protection policy.

The most important 'government strategies' which the Commission can employ is to take to the ECJ member states which have failed to implement correctly Natura 2000 measures. The most important legally binding measures of Natura 2000 are two*directives* (i.e. the 1979 wild birds directive (79/409) and the 1992 habitats directive (92/43). Directives are by far the most commonly used legal measures in EU environmental policy. Importantly, directives first have to be transposed into national legislation before they become applicable in member states. They therefore differ from EU *regulations*, which are directly applicable though not widely used for environmental policy.[2] The strong reliance on directives in the common environmental policy can be explained by the fact that they allow for a certain degree of flexibility during the transposition stage. This should make it easier for member state governments to ensure that EU legislation fits well into the existing body of national environmental legislation. In theory this should minimise the risk of implementation failure of EU environmental legislation. However, transposition failure (i.e. the failure to adopt national legislation on time and/or to transpose correctly the EU directive into national legislation) was so widespread for the 1979 wild birds and the 1992 habitats directives that the Commission threatened, at various points in time, all EU member states with legal action. However, as has been shown by the previous chapters in this volume, despite the legal action taken by the Commission there have been serious transposition delays in many member states. More importantly, as has again been shown by the previous chapters in this volume, the practical implementation (rather than just the formal transposition) of Natura 2000 measures has been very poor in many member states.

However, before we can assess in comparative perspective the most central empirical findings of the implementation process of Natura 2000 measures as put forward by the previous chapters of this volume, it is important to recall briefly the most important justifications (i.e. sources of legitimacy) for the EU's involvement in environmental policy including nature protection.

---

[2] However, the ECJ has ruled that under certain circumstances directives can also become directly applicable thus blurring the distinction between directives and regulations.

20.4 Legitimising the Adoption of EU Environmental Policy Measures

There are four major justifications for the EU's involvement in environmental policy. First, many environmental problems are transnational problems which cannot be solved by member states acting in isolation. The protection of migratory wild birds is a classic example for this.

Second, different national environmental standards (e.g. different car emission limits), can create barriers to trade and therefore endanger the functioning of the EU's internal market which is one of the EU's main objectives. As was discussed above, this justification was the main driver of EU environmental policy measures until the SEA came into force at the beginning of the mature phase in 1987. However, it was of little importance for the EU's early nature protection measures.

Third, many pollution problems are the result of legitimate human activities (e.g. industrial production) which can lead to 'negative externalities' such as pollution, the extinction of bird species and the destruction of natural habitats. The conventional neo-liberal and socialist views were that there is a trade off between environmental protection and economic growth. However, these views were challenged by advocates of the ecological modernisation doctrine who argued that stringent environmental regulations would benefit both the environment and the economy (Hajer 1995; Jänicke and Jacob 2006; Weale 1992). Proponents of the ecological modernisation doctrine gained considerable support in some member states (such as Denmark, Germany and the Netherlands) and amongst some EU actors (such as the EP and DG Environment) during the late 1980s and early 1990s. However, in recent years sustainable development, which puts equal weight on environmental, economic and social considerations and advocates wide stakeholder participation, has overtaken the ecological modernisation doctrine as a paradigm for EU environmental policy. However, the economic recession in Europe during the early 1990s and the debate about the principle of subsidiarity in the wake of the 'No' vote in the first Danish referendum on the Maastricht Treaty had already pushed onto the defensive proponents of ecological modernisation.

Fourth, from the late 1960s onwards, it became increasingly clear that environmental degradation is incompatible with 'the raising of living standards' which is one of the EU's primary objectives. Since 1972, the EU has adopted environmental policy measures both to protect the environment and to raise the 'quality of life' for its citizens. From its early beginnings, EU environmental policy (including nature protection) was therefore also a means of lending additional legitimacy to the European integration project which had stalled in the late 1960s (Wurzel 2007). This line of argument is supported by the fact that the first nature protection measures (e.g. the 1979 birds directive (79/409)) were adopted even before the SEA introduced explicit environmental Treaty competences in 1987.

20.5 Reassessing the Main Themes

The importance of the following five main themes stands out from the contributions to this volume: (1) policy and/or state failure; (2) legitimacy; (3) the notion of trust; (4) scientific versus local knowledge; and, (5) political bargaining versus deliberation. However, the following three important research questions for nature protection policy in Europe have either not been addressed at all or barely touched in the previous chapters by the contributors to this volume: (1) Does the EU present the optimal regulatory arena for nature protection policy?; (2) Can a 'North–South' and/or a 'new–old' member state dimension be identified in EU nature protection policy?; and, (3) Has the integration of nature protection requirements into other policy sectors made any progress on the EU and/or member state level? This concluding chapter will not be able to give a conclusive answer to these under-researched questions. However, it is worth flagging up their importance in order to get a fuller picture of the achievements and failures of EU nature protection policy. It should also throw some light on the possible revision of Natura 2000 measures.

20.5.1 Main Themes Addressed in the Contributions to this Volume

20.5.1.1 Policy and/or State Failure

The EU's 1979 wild birds directive (79/409) and 1992 habitats directive (92/43) were, to a significant degree, triggered by (member) states' failure to adopt and implement adequate policies for the protection of wild birds and habitats. This is unsurprising as the protection of wild birds and transnational habitats requires collective action between (member) states. Most member states had already in place national nature protection policies before they joined the EU. For example, Susana Aguilar Fernández has pointed out in chapter 6 that Spain adopted its first major nature protection measures already in 1918. There were also a considerable number of international agreements in the field of nature protection such as the Bern convention which came into force in 1982 (see, for example, chapter 4 by Rauschmayer and Behrens). However, while national action was too narrow in terms of its focus and instrument reach, international action often produced little more than lowest common denominator agreements between states which had very different interests and traditions as regards the protection of wild birds and habitats. Moreover, most international environmental agreements suffered from serious implementation deficits as it was difficult to enforce international law (Victor et al. 1998). EU nature protection legislation, which allowed recourse to the ECJ in order to enforce supranational laws in a top down manner, was therefore perceived by member states as a promising solution to avoiding policy failure and/or state failure. However, as the contributions to this volume have shown, Natura 2000 also suffered from serious implementation problems although overall they tend to be less serious compared to the implementation of international nature agreements

(Victor et al. 1998; see also chapter 10 by Chmielewski and Krogulec and chapter 4 by Rauschmayer and Behrens).

A recent assessment of Europe's environment by the European Environment Agency (EEA), concluded that although some progress had been made, the threat to biodiversity in general and wild birds and natural habitats in particular was still a serious one (EEA 2003: 230–49). The EEA criticised that the focus of past nature protection measures 'has long been on the most threatened and flagship species, such as large carnivores, and the population trends for these vary considerably. However, some previously common species are now facing serious decline towards very unstable populations and reduced distribution ranges' (EEA 2003: 240). Particularly the findings presented in chapter 2 by Leistra, Keulartz and Engelen and in chapter 4 by Rauschmayer and Behrens support the EEA's warning. In a speech in 2006, the Environmental Commissioner Stavros Dimas made a dramatic appeal for increased efforts in order to avoid nature protection policy failures when he stated that '[s]topping the loss of biodiversity and limiting climate change are the two most important challenges facing the planet. In one fundamental way biodiversity loss is an even more serious threat, because the degradation of ecosystems often reaches the point of no return – and because extinction is forever' (*ENDS Europe Daily* 30 May 2006).

One of the most visible nature protection policy disasters has been the breaking of a dam which allowed toxic sludge to flood parts of the nearby protected nature reserve Doñana, in Andalusia (Spain) in 1998. Sudden and widely visible nature protection policy disasters attract great media attention and environmental group activism, thus opening up policy windows for change as was the case in Spain. However, as most of the contributions to this volume have shown, failures by the state (or regional authorities) to protect nature typically become visible only gradually and over a relatively long period of time. In this case, local residents (see chapter 14 on Flanders by Bogaert and Leroy and chapter 12 on the North Yorkshire Moors by Leroy Oughton and Wheelock) and (local) environmental groups (see chapter 16 on France by Pinton, chapter 18 on Germany by Leibenath, and chapter 10 on Poland by Chmielewski and Krogulec) often play a crucial role in bringing about policy change.

A reoccurring theme in most of the chapter contributions to this volume is that one of the deficiencies from which the early output oriented national, international and EU nature protection policies suffered was that they were devised by a relatively small elite which frequently failed to consult stakeholders, including local residents and/or policy actors who had to implement the adopted policy measures on the ground. All of the chapter authors in this volume seem to agree that top down policy approaches to nature protection carry a high risk of policy and/or state failure.

20.5.1.2 Legitimacy

The introductory chapter by Engelen, Leistra and Keulartz draws on the analytically very useful distinction between output, input and through-put legitimacy which has been taken up (either explicitly or implicitly) by most chapter authors in this volume.

The increasing importance of input and through-put legitimacy including the wide consultation of stakeholders, transparent procedures and clear lines of accountability have been emphasised, although to varying degrees, by all chapter authors in this volume. The EU's recent shift of emphasis away from output oriented detailed substantive policy measures, which are implemented in a top down fashion, towards procedural measures and non-regulatory policy instruments (such as benchmarking) seem to confirm the alleged rise of new modes of governance and the simultaneous decline of 'command-and-control' regulation which has been the most widely used policy tool for traditional government. As was explained above, the Commission's *White Paper on European Governance* has, so far, given the clearest expression in favour of new modes of governance on the EU level (CEC 2001).

However, EU nature protection policy is (still) largely conducted with the help of traditional top down regulatory tools such as the wild birds directive (79/409) and the habitats directive (92/43). A classic example for the top down character of the implementation process of Natura 2000 measures is the so-called Leybucht decision (case 57/89) by the ECJ. The Commission took Germany to the ECJ for breaching the habitats directive after local authorities in Germany authorised the building of a dyke as part of a coastal defence project. The ECJ ruled that member states had no general power to modify or reduce specially protected areas (SPAs) once they had been declared. It argued that member states are allowed to undertake modifications only in exceptional circumstances while 'general interests superior to the ecological objective envisaged in the Directive' are at stake (Haigh 2005: 9.9). The Leybucht case illustrates well the difference between EU environmental legislation and international environmental laws. Clearly, under international environmental law it would have been much more difficult to enforce in a top down manner the protection measures put forward in the habitats directive.

However, top down enforcement of EU laws also brings about disadvantages as has been pointed out by many chapter authors in this volume. For example, chapter 16 by Pinton explains that the level of local resistance in France to the EU's Natura 2000 measures was at least partly dependent on the level of stakeholder inclusion (ie. Low levels of consultation and representation triggered high levels of resistance) during the implementation phase. Similarly, Björkell argues in chapter 8 that a bottom up approach, which allowed for greater participation and relied on transparent procedures, produced better implementation results on the local level in Finland whereas Natura 2000 measures have caused the largest amount of environmental conflict in recent times.

However, the inclusion of a wide range of stakeholders in the policy-making process is usually costly and time-consuming. It therefore may delay the urgently needed adoption of policy measures such as protection measures for species which are threatened by extinction. Importantly, stakeholder consultation and representation may not reduce significantly the implementation deficit (from which EU environmental policy in general and Natura 2000 measures in particular suffer) if it is limited to the implementation phase (i.e. the post-decisional phase). According to many of the chapter authors in this volume, one of the main reasons for the generally disappointing results of increased input and through-put legitimacy during the implementation phase is due to the lack of trust between elite EU and national decision makers, who have adopted the EU Natura 2000 measures without much stakeholder consultation, and regional/local stakeholders who have to implement these measures.

20.5.1.3 Trust

Most chapter authors in this volume have argued that a high degree of trust between policy actors can facilitate the implementation of nature protection policy measures. However, as, for example, Björkell has pointed out in chapter 8, trust can be more easily established on the local level, where it is less difficult to establish consultation fora in which a relatively small number of policy actors has the opportunity to get to know each other, than on the national and/or supranational level where, amongst others, the large number of policy actors involved in the consultation process often makes it difficult to establish trusted relationships. Moreover, it is not easy to establish a strong notion of trust during the post-decision phase (i.e. the implementation phase) of the policy-making cycle which is usually made up of the following phases: (1) agenda setting; (2) decision-making; (3) implementation; and, (4) revision (see Wurzel 2002 for a more detailed assessment and additional literature). As contributions to this volume have shown, a large number of local actors in many member states have generally remained suspicious of wide consultation efforts by the EU and/or national governments about Natura 2000 measures during the implementation phase because they were not consulted during the decision-making phase (i.e. when the wild birds directive (79/409) and the habitats directive (92/43) were adopted). The revision phase of the Natura 2000 measures therefore offers an opportunity to allow for greater participation and involvement of a wide range of stakeholders.

EU environmental policy in general and nature protection policy in particular would greatly benefit from increased involvement of a wide range of stakeholders who are often in possession of local knowledge which can be crucial for achieving the desired policy objectives.

20.5.1.4 Scientific Versus Local Knowledge

However, as several of the chapters in this volume have shown (see, for example, chapter 14 by Bogaert and Leroy, chapter 8 by Björkell, chapter 2 Leistra, Keulartz and Engelen and chapter 4 by Rauschmayer and Behrens) local knowledge may not pull policy makers in the same direction as scientific knowledge. EU environmental policy has heavily relied on scientific knowledge during the agenda setting and decision-making phases. Local knowledge by, for example, environmental groups has played an important role mainly during the implementation phase of the EU environmental policy-making cycle. In chapter 10, Chmielewski and Krogulec have pointed out that Polish environmental groups published so-called shadow lists which showed that the government had designated only a fraction of the SPAs which fall within the scope of the Natura 2000 measures. The Polish government subsequently submitted to the Commission a list with a much higher number of SPAs. As was pointed out above, environmental groups often take on a whistleblowing function when they inform the Commission about member state implementation failures.

Chapter 8 by Björkell and chapter 12 by Oughton and Wheelock have also shown the importance of local knowledge for the bottom up management of local commons. Local traditions and conservation practices have made important contributions towards the conservation of nature protection sites such as the Bergö-Malax Outer Archipelago in Finland (chapter 8) and the North Yorkshire Moors in the United Kingdom (chapter 12). However, there is no guarantee that stakeholder involvement and local knowledge will lead to successful conservation policies.

The limitations of scientific knowledge in nature protection policy and the importance of conflicting nature protection and other societal interests were well illustrated in chapter 2 on wintering geese in the Netherlands by Leistra, Keulartz and Engelen and in chapter 4 on the great cormorant by Rauschmayer and Behrens. Both chapters showed that nature protection measures for threatened species may trigger unintended negative consequences for other species and/or legitimate societal interests such as fishing or farming.

20.5.1.5 Political Bargaining Versus Deliberation

How to reconcile differences in member state and/or societal interests and conflicting ideas (including scientific and local knowledge) about the most appropriate nature protection measures constitutes a major challenge for the EU's multi-level decision-making process. However, especially chapter 18 on Germany by Leibenath and chapter 6 on Spain by Fernández have shown that federal systems (Germany) and quasi-federal systems (Spain) encourage political bargaining between different levels of government which may hinder the adoption and/or implementation of nature protection measures. On the other hand, federal and quasi-federal systems may react more quickly to nature protection policy challenges that have emerged on the local or regional level. The Müritz park (see chapter 18 by Leibenath) and the nature reserve Doñana in Andalusia (see chapter 6 by Fernández) are good examples which support this claim.

Political bargaining rather than problem solving is often said to be the dominant feature of decision-making on the EU level (Scharpf 1988; 1993). However, compared to the international level, the EU provides a surprising number of access points to stakeholders. For example, the Commission has traditionally relied heavily on the environmental groups to find out about implementation failures in EU nature protection policy. Nowadays it also tends to consult a wide range of stakeholders before publishing its policy proposals although this has not always been the case in the past. Moreover, the EP is very much open to lobbying from stakeholders while Members of the EP (MEP) have direct links to constituencies directly affected by EU nature protection measures. Finally, the Environmental Council, which together with the EP adopts EU environmental policy measures, has shown a remarkable readiness to adopt also problem-solving strategies (see, for example, Scharpf 1996).

This is not to deny that political bargaining about national interests is an important feature for all Council of Minister negotiations.

The EU has relied heavily on top down political bargaining strategies during the implementation phase of the Natura 2000 measures. Taking member states to the ECJ and threatening them with withholding funds from the EU structural funds if they persistently fail to implement the Natura 2000 measures are good examples for top down political bargaining strategies. Much of the empirical evidence put forward in this volume suggests that the EU Natura 2000 measures have made a difference in terms of increased nature protection. However, almost all of the chapter authors, although for different reasons, seem to suggest that traditional top down political bargaining and/or 'command-and-control' regulations may have run their course. The importance of consensual deliberative decision-making procedures which include a wide range of stakeholders while paying attention to procedural fairness, transparency and accountability has been raised by almost all of the chapter authors in this volume. It will be a major challenge for multi-level EU nature protection policy to move away from traditional ('command-and-control) regulations towards more deliberative modes of governance. However, whether these new modes of governance will be able to reduce the implementation deficit which seems to be endemic in (EU, national and international) nature protection policy remains to be seen.

20.5.2 Under-researched Themes

Before putting forward a brief conclusion it is worth pausing for a moment in order to consider important questions mentioned which were under-researched in the chapters in this volume. The intention is not to provide in Sherlock Holmes fashion an answer to 'why the dog did not bark' but rather to draw the attention of the reader towards themes which currently play an important role in EU environmental policy and to flag up issues for possible future research.

20.5.2.1 The EU as the Optimal Regulatory Arena

None of the authors explicitly posed the research question of whether the EU represents the optimal regulatory arena for nature protection measures. This was arguably due to the country and/or case study focus of the chapters which took EU and/or member state nature protection measures as the given starting point for the empirical research. Several authors (for example, Chmielewski and Krogulec in chapter 10) referred to international nature protection agreements such as the above mentioned Bern convention. However, with the possible exception of chapter 4 by Rauschmayer and Behrens, no comparison was undertaken as regards the implementation of the EU's Natura 2000 measures and international nature protection measures. Policy makers often have to decide between relatively weak international environmental agreements, which tend to offer a better geographic reach for nature protection measures than EU measures, and more effective (in terms of output legitimacy) supranational regulatory tools that are available only on the EU level.

The wider EU environmental policy literature and international environmental politics literature seems to suggest that EU environmental policy suffers from an implementation deficit which is somewhat less serious than international environmental agreements (Victor et al. 1998; Wurzel 2002). In terms of output legitimacy one could therefore argue that EU environmental policy has achieved a higher degree of legitimacy compared to international environmental agreements. However, as was pointed out by almost all of the chapter authors in this volume, the EU's nature protection policy has to a large degree relied on traditional 'command-and-control' regulation and top down implementation strategies which have triggered considerable resistance on the local and regional level. Clearly the EU offers an attractive regulatory arena for those policy actors who favour traditional regulatory instruments on a supranational scale. Whether the EU will remain as attractive a regulatory arena if new modes of governance become more popular for nature protection policies remains to be seen. Moreover, future enlargements may trigger the need for differentiated nature protection standards in order to take into account both the difference in species and habitats as well as the level of economic developments (Holzinger and Knoepfel 2000).

20.5.2.2 'North–South' and/or a 'New–Old' Member State Dimension

The question of whether there is a 'North–South' and/or a 'new–old' member state dimension in EU nature protection policy was not addressed systematically in any of the chapters in this volume. This is no surprise as all of them focused on one particular country with the exception of chapter 4 on the great cormorant by Rauschmayer and Behrens who flagged up some differences in policy between Denmark, France and Italy. The general EU environmental policy literature has often pointed out that within the EU the 'environmental pioneers' (Andersen and Liefferink 1997) are mainly from the Northern member states (and include Denmark, Germany and the Netherlands and, since they have joined the EU in 1995, Austria, Finland and Sweden). There is also widespread agreement that the 'new' Central and Eastern European member states are generally faced with graver environmental problems while having fewer financial resources to tackle environmental pollution (Holzinger and Knoepfel 2000).

However, the empirical evidence put forward in the chapters in this volume suggests that environmental leader states such as Finland (see chapter 8 by Björkell) and Germany (see chapter 18 by Leibenath) also had great difficulties in implementing correctly the EU's Natura 2000 measures. On the other hand, chapter 6 by Fernández shows that Spain had already adopted some important nature protection measures before joining the EU. Chapter 10 by Chmielewski and Krogulec shows that Poland has made major strides towards the protection of wild birds and habitats since joining the EU. However, important national nature protection measures were already in place long before Poland applied for EU membership. On the empirical basis of the chapters in this volume, it would be wrong to conclude that there is no 'North–South' and/or a 'new–old' member state dimension in EU nature protection policy. However, it would certainly be worth investigating whether it is as significant as some of the general EU environmental policy literature seems to suggest.

20.5.2.3 Integrating Nature Protection Requirements into Other Policy Sectors

The integration of nature protection requirements into other policy sectors (such as agriculture, transport and economic) clearly plays an important role for the success or failure of the EU's Natura 2000 measures. Yet, most of the chapters in this volume remain silent about the need to better integrate nature protection requirements into other policy sectors. To be fair, several chapters have highlighted the existence of conflicting interests between different policy actors. For example, chapter 18 by Leibenath points to conflicts between nature protection and regional development while chapter 4 by Rauschmayer and Behrens flags up conflicts between nature protection goals and the interests of fishermen. However, it is doubtful whether increased deliberative policy-making, which seems to be the favoured policy-making mode put forward by several chapter authors, will be able to solve the problem that nature protection requirements are not taken into account in other policy sectors.

20.6 Conclusion

The 1979 wild birds directive (79/409) and the 1992 habitats directive (92/43) together with the LIFE nature fund are the core measures of Natura 2000 which constitutes the cornerstone of EU nature protection policy and its more recent biodiversity strategy. As contributions to this volume have shown, the EU's Natura 2000 measures have had a considerable impact on member states' nature protection policies. In terms of output legitimacy, it is fair to say that Natura 2000 has contributed towards improvement of the environment in Europe, although major threats to endangered species and the environment still exist (EEA 2003).

However, the EU's nature protection policy measures have often lacked input and throughput legitimacy, as was illustrated by most chapter authors. The fact that the 1979 wild birds directive was introduced even prior to the EU having explicit environmental treaty provisions, supports well Jack Hayward's (1996) claim that European integration has, up to now, largely been achieved through 'integration by stealth'. The EU's nature protection policy has essentially used a top down traditional regulatory approach. There have been attempts to make use of 'smart regulation' (Gunningham and Grabosky 1998). For example, cost effectiveness considerations in EU nature protection measures have involved not only calculations of short-term costs but also the medium to long term benefits. However, new modes of governance, which are generally associated with wider participation and consultation with stakeholders, have barely been used in EU nature protection policy although they have been tested in other areas of EU environmental policy. When trying to improve the implementation of Natura 2000 measures the EU has traditionally relied very heavily on top down 'government strategies'. Bottom up 'governance strategies' (such as the wider consultation of stakeholders) are more recent measures whose use has not (yet) been widespread in nature protection policy.

The EU's nature protection policy therefore does not provide a good illustration for the claim by Rosenau and Czempiel (1992) that there has been a shift towards 'governance without government' in the international arena. This is surprising as EU and national nature protection policies are highly internationalised policies which have been greatly affected by developments on the international level. EU nature protection policy has been characterised by a high degree of traditional government regulation. The empirical evidence presented in the chapters in this volume has uncovered the deficiencies of an over-reliance on output legitimacy. However, most chapter authors have found it more difficult to put forward alternative (prescriptive) solutions to the problem of deficiencies in the input and through-put legitimacy of EU nature protection policy.

Acknowledgments The author would like to thank the editors and organisers of a workshop on which this book is based, Ewald Engelen, Gilbert Leistra and Jozef Keulartz for their assistance and efforts. He is grateful to the Hull Environmental Research Institute (HERI) at the University of Hull which has funded part of the research on which this chapter is based.

References

Andersen M.S., and D. Liefferink, eds. (1997), *European Environmental Policy. The Pioneers*, Manchester: Manchester University Press.

Baacke, R. (2000). 'Forentwicklung des Umweltrechtes – Europa als Modell oder Problem?', Speech by the *Staatssekretär* for the Environmental Ministry at the Humboldt University of Berlin on 31 May 2000.

Börzel, T. (2002). 'Pace-setting, Foot-Dragging, and Fence-Sitting: Member State Responses to Europeanization'. *Journal of Common Market Studies*, 40(2): 192–214.

Bungarten, H. (1978). *Umweltpolitik in Westeuropa*. Bonn: Europa Union Verlag.

CEC (1973). *Programme of Action of the European Communities on the Environment*. OJ No C 112, 20 December.

CEC (1977). *European Community Action Programme on the Environment (1977 to 1981)*. OJ C, 13 June.

CEC (1987). *European Community Policy and Action Programme on the Environment (1987–1992)*. OJ No C 70, 18 March.

CEC (1992). *Environmental Legislation. Volume 4. Nature*. Brussels: Commission of the European Communities.

CEC (1996). *European Community Environmental Legislation. Volume 4 – Nature*, Brussels: Commission of the European Communities.

CEC (2001), *European Governance. A White Paper. COM (2001) 428 final*, Brussels: Commission of the European Communities.

CEC (2006). *Communication from the Commission. Halting the Loss of Biodiversity by 2010 – and Beyond. Sustaining Ecosystem Services for Human Well-Being. COM (2006), 216 final*, Brussels: Commission of the European Communities.

EEA (2003). *Europe's Environment: the Third Assessment, Environmental Assessment Report No. 10*. Copenhagen: European Environment Agency.

EEPL (various years). European Environment & Packaging Law Weekly; Tunbridge Wells: Agra Informa Ltd.

ENDS Europe Daily (various years). E-mail service, London: Environmental Data Services.

Finer, S. (1970). *Comparative Government*. Harmondsworth: Penguin.

Gunningham, N., and Grabosky, P. (1998). *Smart Regulation*. Clarendon: Oxford.

Haigh, N. (2005). *Manual of Environmental Policy*. Leeds: Maney Publishing.

Hajer, M. (1995). *The Politics of Environmental Discourse. Ecological Modernization and the Policy Process*. Oxford: Oxford University Press.

Hayward, J. (1996). 'Conclusion: Has European Unification by Stealth a Future?' In J. Hayward (ed.), *Elitism, Populism, and European Politics*. Oxford: Clarendon Press Oxford, pp. 252–58.

Héritier, A., C. Knill and S. Mingers (1996). *Ringing the Changes in Europe*. Berlin: Walter de Gruyter.

Hildebrand, P. (1992). 'The European Community's Environmental Policy, 1957–1992'. *Environmental Politics* 194: 13–44.

Holzinger K., and P. Knoepfel, eds. (2000). *Environmental Policy in a European Union of Variable Geometry? The Challenge of the Next Enlargement*. Basel, Genf, Munich:

Inglehart, R. (1971). 'The Silent Revolution in Europe: Intergenerational Change in Post-Industrial Societies'. *American Political Science Review* 65: 991–1017.

Jänicke, M., and K. Jacob, eds. (2006). *Environmental Governance in Global Perspective. New Approaches to Ecological and Political Modernisation*, Berlin: Freie Universität Berlin.

Jordan, A., R. Wurzel and A. Zito, eds. (2003). *New Instruments of Environmental Governance*. London: Frank Cass.

Jordan, A., R. Wurzel and A. Zito (2005). 'The Rise of "New" Policy Instruments in Comparative Perspective: Has Governance Eclipsed Government?' *Political Studies* 53(3): 477–96.

Jordan, A., R. Wurzel and A. Zito (2006). 'Policy Instrument Innovation in the European Union'. In G. Winter, ed. *The Multi-level Governance of Global Environmental Change*, Cambridge: Cambridge University Press, pp. 470–91.

Kloepfer, M. (1998). *Umweltrecht*. Munich: Beck.

Knill, C. (2001). *The Europeanisation of National Administrations*. Cambridge: Cambridge University Press.

Knill, C., and A. Lenschow, eds. (2000). *Implementing EU Environmental Policy*. Manchester: Manchester University Press.

Mazey, S., and J. Richardson, eds. (1993). *Lobbying in the European Community*. Oxford: Oxford University Press.

Osborne, D., and T. Gaebler (2001). *Re-inventing Government*. Reading, Mass: Addison Wesley.

Rehbinder E., and R. Stewart (1985). *Integration Through Law. Europe and the American Federal Experience*. Berlin: Walter de Gruyter.

Rosenau, J., and E.-O. Czempiel, eds. (1992). *Governance without Goverment: Order and Change in World Politics*. Cambridge: Cambridge University Press.

Sabatier, P. (1993). 'Policy Change over a Decade or More'. In P. Sabatier and H. Jenkins-Smith, eds. *Policy Change and Learning. An Advocacy Coalition Approach*. Boulder: Westview Press, pp. 13–41.

Scharpf, F. (1988). 'The Joint-Decision Trap: Lessons from German Federalism and European Integration'. *Public Administration* 66: 239–78.

Scharpf, F., ed. (1993). *Games in Hierarchies and Networks. Analytical and Empirical Approaches to the Study of Governance Institutions*. Frankfurt: Campus.

Scharpf, F. (1996). 'Negative and Positive Integration in the Political Economy of European Welfare States'. In G. Marks et al., eds. *Governance in the European Union*. London: Sage, pp. 15–39.

SRU (2004). *Umweltgutachten 2004*. Berlin: Der Rat von Sachverständigen für Umweltfragen.

Victor, D., K. Raustiala and E. Skolnikoff, eds. (1998). *The Implementation and Effectiveness of International Environmental Commitments: Theory and Practice*. Cambridge, Massachusetts and London: The MIT Press.

Weale, A. (1992). *The New Politics of Pollution*. Manchester: Manchester University Press.

Weale, A. (1996). 'Environmental Rules and Rule-Making in the European Union'. *Journal of European Public Policy* 3(4): 594–611.

Weale, A., G. Pridham, M. Cini, D. Konstadakopulos, M. Porter and B. Flynn (2000). *Environmental Governance in Europe*. Oxford: Oxford University Press.

Wurzel, R.K.W. (1993). 'Environmental Policy'. In J. Lodge, ed. *The European Community and the Challenge of the Future*, second edition, London: Pinter, pp. 178–99.

Wurzel, R.K.W. (2002). *Environmental Policy-making in Britain, Germany and the European Union*. Manchester: Manchester University Press.

Wurzel, R.K.W. (2003). 'Environmental Policy: An Environmental Leader State under Pressure?' In K. Dyson and K. Goetz, eds. *Germany, Europe, and the Politics of Constraint*. Oxford: Oxford University Press, pp. 289–308.

Wurzel, R.K.W. (2004). 'The Europeanisation of German Environmental Policy: From Environmental Leadership to Partial Mismatch?' In A. Jordan and D. Liefferink, eds. *The Europeanisation of Environmental Policy: European Union and Member State Perspectives*. London: Routledge, pp. 99–117.

Wurzel, R.K.W. (2007). 'Environmental Policy: Actors, Leaders and Laggard States'. In J. Hayward, ed. *Leaderless Europ*. Oxford: Oxford University Press (forthcoming).

The International Library of Environmental, Agricultural and Food Ethics

1. H. Maat: *Science Cultivating Practice*. A History of Agricultural Science in the Netherlands and its Colonies, 1863-1986. 2002   ISBN 1-4020-0113-4
2. M.K. Deblonde: *Economics as a Political Muse*. Philosophical Reflections on the Relevance of Economics for Ecological Policy. 2002 ISBN 1-4020-0165-7
3. J. Keulartz, M. Korthals, M. Schermer, T.E. Swierstra (eds.): *Pragmatist Ethics for a Technological Culture*. 2003   ISBN 1-4020-0987-9
4. N.P. Guehlstorf: *The Political Theories of Risk Analysis*. 2004
       ISBN 1-4020-2881-4
5. M. Korthals: *Before Dinner*. Philosophy and Ethics of Food. 2004
       ISBN 1-4020-2992-6
6. J. Bingen, L. Busch (eds.): *Agricultural Standards*. The Shape of the Global Food and Fiber System. 2005   ISBN 1-4020-3983-2
7. C. Coff: *The Taste of Ethics: An Ethic of Food Consumption*. 2006
       ISBN 1-4020-4553-0
8. C.J. Preston and Wayne Ouderkirk (eds.): *Nature, Value, Duty: Life on Earth with Holmes Rolston III*. 2007   ISBN 1-4020-4877-7
9. Dirk Willem Postma: *Why Care for Nature? In search of an ethical framework for environmental responsibility and education*. 2006 ISBN 978-1-4020-5002-2
10. P.B. Thompson: *Food Biotechnology in Ethical Perspective*. 2007
       ISBN 1-4020-5790-3
11. P.W.B. Phillips, C.B. Onwuekwe (eds.): *Accessing and Sharing the Benefits of the Genomics Revolution*. 2007   ISBN 978-1-4020-5821-9
12. P. Pinstrup-Andersen, P. Sandøe (eds.): *Ethics, Hunger and Globalization*. In Search of Appropriate Policies. 2007   ISBN: 978-1-4020-6130-1
13. H. Zwart: *Understanding Nature*. Case Studies in Comparative Epistemology. 2008   ISBN: 978-1-4020-6491-3
14. J. Keulartz, G. Leistra (eds.): *Legitimacy in European Nature Conservation Policy*. Case Studies in Multilevel Governance. 2008 ISBN: 978-1-4020-6509-5

Printed in the United States
98534LV00002B/271-282/A